Down and Out
in the New Economy

Down and Out

in the

New Economy

How People Find
(or Don't Find)
Work Today

Ilana Gershon

The University of Chicago Press
Chicago and London

The University of Chicago Press, Chicago 60637
The University of Chicago Press, Ltd., London
Published 2017
Paperback edition 2024
Printed in the United States of America

33 32 31 30 29 28 27 26 25 24   1 2 3 4 5

ISBN-13: 978-0-226-45214-2 (cloth)
ISBN-13: 978-0-226-83322-4 (paper)
ISBN-13: 978-0-226-45228-9 (e-book)
DOI: https://doi.org/10.7208/chicago/9780226452289.001.0001

Library of Congress Cataloging-in-Publication Data

Names: Gershon, Ilana, author.
Title: Down and out in the new economy : how people find (or don't find)
    work today / Ilana Gershon.
Description: Chicago ; London : The University of Chicago Press, 2017. |
    Includes bibliographical references and index.
Identifiers: LCCN 2016040351 | ISBN 9780226452142 (cloth : alk. paper) |
    ISBN 9780226452289 (e-book)
Subjects: LCSH: Job hunting—United States. | Employee selection—
    United States. | Online social networks in business—United States. |
    Industrial relations—United States. | United States—Economic
    conditions—2009–
Classification: LCC HF5382.75.U6 G465 2017 | DDC 650.140973—dc23
LC record available at https://lccn.loc.gov/2016040351

♾ This paper meets the requirements of ANSI/NISO Z39.48-1992
(Permanence of Paper).

To Friedrich Hayek.
If his philosophy had not been so influential,
I would never have had to write this book.

Contents

Preface   A Book about Advice, Not an Advice Book

For over a decade, I have been training college students to go out into the world with a bachelor's degree, but before I started writing this book, I didn't know much about how they would find their living. How do people find good job opportunities in the first place? What makes a good resume? I had no answers to these practical questions, and I felt as frustrated and powerless as my students. So I decided to research what it takes to get hired in corporate America these days. If I've learned anything, it's that some of what other people say is good advice, some is very good advice, and some of it is just plain awful. I want to show you how to tell the difference.

I'm not a career counselor. Career counselors provide suggestions to people looking for a wide range of jobs in many different kinds of workplaces. To give advice, they have to offer standardized techniques for people with many different backgrounds and goals. As a result, their advice often centers on telling people how to produce standardized ways of representing themselves as employable — similar resumes, similar business cards, similar LinkedIn profiles, and similar networking strategies. But the more I talked to hiring managers, recruiters, and HR professionals about how they sort through job applications, the more convinced I became that there is no one-size-fits-all answer to the question of how you get a job.

We have reached a point at which standardized advice is making hiring immensely confusing and difficult. Job seekers are using standardized advice and standardized forms to simplify their complex work experiences into easily scanned text that will allow employers to quickly compare what in reality might be very different experiences and backgrounds. All of this is supposed to make it easier for the people doing the hiring. But it turns out that these resumes and interviews don't actually provide enough information to make sensible, well-informed decisions about whom to hire. Amid all this standardization, what people really want to know is what it will be like to work with the applicant. They try a number of techniques to figure this out—some of them are more effective than others, and some clash with the techniques other people, even from the same office, might use. Far from making employment decisions more predictable and rational, standardization can actually make hiring more idiosyncratic and frustrating for everyone involved. And the advice generator continues . . .

I am more interested in ways of breaking through the barriers of standardization that separate job seekers from the people doing the hiring. Job seekers need to know how to learn about the specific companies they want to join. They also need to realize that when they apply for a job, they are dealing with a group of people struggling together to interpret a pile of slightly too similar documents and applicants who are expected to say the same kind of thing over and over again. In short, once job seekers have figured out how to have a good enough resume, and how to be a good enough interviewee—what career counselors are often very good at explaining—they also need techniques for figuring out how different workplaces function and whether they in fact want to join them.

I also want to get to the root of the technology problem in hiring and figure out why all the new technologies people use these days inspire so much conflicting advice. All this advice is sparked not just by the uncertainty of the job market; it is also a result of people's uncertainty about how to use new technologies. Every new medium changes how people communicate, however slightly. For example, what you communicate when sending a text is different from what you communicate when posting a Facebook update, even if the words are identical. And emailed information circulates differently than information sent in a message to a LinkedIn group. When job seekers are applying for a job, they tend to focus on the forms. They hope that if they just compose the right resume, or the right LinkedIn profile, they will get the job. These forms seem like they might possibly be under your control, and, let's face it, not much else does when you are applying for jobs. But what might have seemed under your control on paper is now a bit mystifying on-screen. This uncertainty sparks advice, and lots of it. But sometimes that advice only reveals how the advice giver would like things to be, not necessarily how things are. The rules are still very much in the making.

We are not all equal players in the new employment game. Over the past thirty years, the employment contract and what it means to be a worker in general have changed. Workers are living anxious, unstable lives in which they are told they must shoulder most of the responsibility for changes in today's economy. Much of the advice out there makes things worse. It benefits companies far more than it benefits workers. Lots of recommendations — to brand your self, to be passionate about work, to present your self as a bundle of temporary business solutions for market-based problems — tell people to just accept and adapt to an uncertain and unstable world. Instead of being a craftsperson, developing and honing a set of skills over time, you are expected to be more of a jack-of-all-trades, developing a range of skills that might be useful in an unknown and ever-changing marketplace.[1] Instead of company loyalty, you are now expected to feel passion for your vocation and to be driven to prioritize work over all other obligations. This advice is given to people who are already feeling vulnerable — anxious about how they will make a living and willing to adopt any suggestion that will help them get a job. This is the kind of advice I am opposed to.

We don't need more and more techniques for achieving the impossible, for being the universally desirable candidate for all businesses. We need to understand how we got here, where all this advice is coming from, and how we might make hiring and working more satisfying in this rapidly changing environment. If, at the end of this book, you can see more clearly the challenges of job searching and you can make more thoughtful employment decisions, then I will have achieved my aim. But this book is also an effort to contribute to larger conversations about what it means to work in the first place. It's a conversation that, for reasons you will see in the pages that follow, we all must join.

Introduction **The Company You Keep**

Everything about job searching has changed. At least, that's what I
was told—over and over again—while researching hiring in California's Bay Area. I would hear this, and it would make me pause. After
all, in corporate America, you still have to submit a job application
for an opening, you still have to send in your resume, and you still
have to be interviewed. Yes, it is true that nowadays almost all job
applications are online, and sometimes a recruiter or a hiring manager might look at a LinkedIn profile instead of a resume. For people
who use computers all the time, is an online job application so fundamentally different from a paper one? Is a LinkedIn profile really
that different from a resume? Nothing all that substantive seems to
have changed about the actual forms and interactions necessary for
getting a job. So why were people telling me otherwise?

The shift that people keep telling me about is not in the technology
or the forms people use to apply for jobs, but in how people understand working these days. When people say everything has changed,
they are pointing to a shift in how Americans think of the employer-
employee relationship. People used to think that as employees they
brought capacities and skills to the workplace for a defined period
of time for their side of the bargain. The metaphor that underpinned
this way of understanding being employed was about owning property. That is, workers used to believe that they owned themselves as

though they were property that could be rented to an employer for a certain period of time, with all that implied.[1]

In today's business world, people are no longer supposed to think of the capacities they possess and are willing to sell. Instead they are supposed to think about the services they can provide to a company. The metaphor has changed. People now think they own themselves as though they are businesses—bundles of skills, assets, qualities, experiences, and relationships, bundles that must be consciously managed and constantly enhanced. Popular business literature encourages people to think of themselves in such terms: as the CEO of Me Inc. To be employable you must represent yourself as a business of one, willing to *temporarily* assist other, larger businesses. From this perspective, hiring resembles a business-to-business contract, a short-term connection for solving market-specific problems. This has not been an easy transition, as many job seekers have found out.[2]

Metaphors matter. They are frameworks that strongly influence how you classify and analyze your experiences. Employment metaphors help us decide the proper relationship between employer and employee, how to improve this relationship, and the strategies that should be used when someone wants to change jobs or a company wants to change working conditions. These metaphors are vague and often contextually and historically specific. So when people think of themselves as property, for example, what do they have in mind—a field of wheat, a car, or a suburban house? And when they think of themselves as businesses, do they have in mind start-ups, mom-and-pop stores, or multinational corporations? These metaphors work in part because they *are* vague. Many people can agree about the general implications of the metaphors without actually agreeing on the precise details. And, most importantly, the metaphors are vague enough that they can imply different things in different contexts and still make sense to many people.

The shift in the way we think about employment has been taking place in the United States since the early 1980s, but there was no single moment in which, one morning, everyone in San Francisco woke up thinking "I'm a business now." Even today, not everyone believes in the metaphor of self-as-business, not everyone likes what it means, and not everyone knows that this is how you are supposed to act. But I was talking to people who were looking for work or looking to hire people in the knowledge economy, and for them the metaphor is inescapable.[3] Job seekers were constantly being told by career counselors, motivational speakers, journalists, and each other that this way of thinking is essential in today's corporate job market, a job market filled with temporary jobs.[4] Nowadays, learning how to hire or be hired means learning how to operate as though you are a business, whether you want to or not.

So what exactly happens when you believe that a person can be a business? How does that change hiring, and what kind of dilemmas does it create? To answer these questions, we have to step back and see how the self-as-property metaphor developed in the first place and how it eventually gave way to Me Inc.

Self as Property

Since the philosopher John Locke, people have talked about the self as property.[5] In this metaphor, you "rent" yourself to an employer for a limited period of time, getting yourself back, so to speak, at the end of the day. Why would Locke and other scholars want to turn to property as the key metaphor? In part, they were wrestling with an idea from ancient Greek times—that to be a truly free citizen you had to own property.[6] Influenced by this thought, some political philosophers believed that owning a part of the nation, however small, meant you had a concrete interest in a republic as a whole, and thus property ownership was fundamental to citizenship.[7] During the Industrial Revolution, people were forced to rethink this assumption, especially because a growing majority of men did not own "real" property as the economy became increasingly based on manufacturing. Political thinkers began to transform their notions of property to accommodate their new social and economic conditions.

Hobbes and Locke were interested in why people, if they are fundamentally created free, would ever agree to have a king or a boss. The short answer is that kings, bosses, and other hierarchical relationships often accompany certain forms of security. People are willing to trade some of their autonomy for security. This trade occurs in a market context, according to C. B. MacPherson, an influential political scientist. As he explains, according to these philosophers, people agree to interact with each other based on market exchanges that emerge from the fact that first and foremost they own themselves. They create a market in which they are able to sell their labor, and in return they will receive money, which provides the ability to purchase what others have produced. In a more general sense, when they enter into a social contract (with an employer or with a nation), they receive a certain amount of protection from violence or from crises in exchange for giving up a certain amount of their freedom. Quite simply, they start doing what other people want in exchange for money and stability. Political thinkers and actors in the seventeenth and eighteenth centuries were using the tools and metaphors available to them to solve their social and political problems.[8] The self-as-property metaphor has influenced employment laws and workplace infrastructures ever since.

The definition of property has not stayed the same since the seventeenth century, however, and the changes affect how this metaphor has worked historically. Sometimes, but not always, property was understood as fundamentally a thing that you have, a commodity that you could bring to the market. Thus your capacity to work, which you owned, could also be sold on the market. Over time, people stopped understanding property as a unified thing and increasingly saw it as a bundle of rights—for example, you could have the right to use different parts of the land, such as mineral rights or rights to build on the land. So too with what it meant to be a worker—you didn't sell yourself lock, stock, and barrel to someone else (and become an indentured servant or slave). Instead you sold your capacity to work for a limited period of time during the day.

US historian Amy Stanley points out that one of the major transformations in how people understood the employment contract in the United States occurred in the Reconstruction period, after the Civil War and the end of slavery. Americans had to stop thinking about some people as property that could be owned entirely by others and had to start thinking about everyone's labor as something to be bought. As long as slavery was a tangible possibility, Americans would argue over the precise nature of the difference between being a wage laborer and being a slave. Some labor spokespeople thought there was no real difference, that people who didn't own land themselves were forced to sell their labor to others just to have food on the table and a roof over their heads. Stanley explains: "They claimed not only that wage slaves were unable to sell labor time apart from their persons, but that the sale—to one master or another—lasted for the entire length of their lives. As the long hours of single days stretched on for weeks and years, in perpetuity, the hireling's status edged closer to the slave's."[9]

In part to counter these kinds of arguments, the idea of a work contract became more and more important. At this historical moment, when a worker entered freely into a contract with an employer, it signaled a number of things. First, a contract presupposed the Lockean model of ownership, that people owned themselves prior to the contract and were agreeing to "rent" parts of themselves. Second, the contract also presumed that there were some parts of you that were inalienable, that your labor was not synonymous with your self. So you could sell your labor and retain possession of your person. In addition, your ability to enter into these contracts was a sign that you were free. Your labor was a different kind of commodity than a slave was, although how to make this distinction logically and effectively sparked many arguments.

As slavery receded into history, people developed different historically specific ways of interpreting what it meant to own yourself, or what was significant about a work contract. Yet the metaphor of the self-as-rentable-property continued to provide the conceptual scaffolding for determining what aspects of an employment relationship were available for debate and legislation. Many of the hotly contested moments in US labor history in the nineteenth and twentieth centuries involved working out the parameters and implications of this metaphor. And vestiges of this metaphor still linger on in contemporary US laws, work practices, infrastructures, and technologies.

How did the metaphor of self-as-property shape how people understood the employment contract? First, there was an assumption of exclusivity built into this version of an employment contract. Just as you can't rent land to more than one person at a time, so too with work—that is, you shouldn't work for multiple employers at once (and especially not in the same time period). These were the days when moonlighting—working a second job—was something workers generally had to hide from their employers. This form of exclusivity eventually morphed into an ideal of company loyalty, which became particularly important in post–World War II America when labor was in short supply. Employers wanted to make sure that they always had a ready supply of workers, and they did so in part by encouraging employees to feel a sense of commitment and obligation to a company.[10] Corporations would encourage company loyalty differently in white-collar workers than they did in blue-collar workers. White-collar workers were promised stable pay, health insurance, and steady promotions for good performances, but this was all largely a tacit bargain, as Peter Cappelli points out.[11] Blue-collar workers often had to organize through unions to get similar rewards for being reliable. In general, company loyalty was seen as something corporations wanted to encourage in their workers, and workers understood that it was a valuable quality that they brought to the bargaining table.

Second, this metaphor implied a sharp boundary between your work and your personal life. After all, when you quit working, from this perspective, you were getting your "self" back; you were no longer on loan to your employer. Your employer had no rights over your behavior when the workday was done. There are many well-known violations of this—Henry Ford famously tried to impose moral standards on his workers, forbidding them to gamble or drink even when they had stopped working for the day. What is important, however, is that Henry Ford's and others' similar efforts were understood at the time as possible infringements on workers' rights. It was a question open for debate whether your employer should have any say on how you behaved on your own time. Other aspects of the self-as-property work contract also inspired arguments when everyone assumed that there should be a certain type of boundary between work and personal life.

For example, people often wondered when precisely the workday might begin. When exactly was the transition between personal time and company time? Did it occur when workers first arrived on their employer's property or only when they arrived at their workstation? Were workers responsible for the time it took to put on the appropriate attire and protective equipment for work or should employers pay workers for the time it took to put on the uniform or protective gear? If you're a police officer, you shouldn't go to a riot without your riot gear. If you are working in a slaughterhouse, you need protective gear to keep the guts and blood from getting all over your personal clothes. In either case, you wouldn't be putting the gear on if you weren't in that job. But on the other hand, you aren't ready for work until you are dressed properly. These types of questions have been at the heart of hotly debated legislation such as the Portal-to-Portal Act of 1947 and the Fair Labor Standards Act of 1938. These acts, passed within ten years of each other, came down on different sides of employer-versus-employee responsibility in debates about the personal-work boundary.[12] While these acts may have taken opposing directions within years of each other, I am more interested in the fact that the question in each case was the same: when precisely did the employee start "renting" him- or herself to the employer?

Indeed, while people might have rented themselves from nine to five to an employer, how people thought about this time eventually became a source of debate between unions and companies. Over time, unions were able to make compelling arguments that employees were not only giving over a portion of their day but were also giving over a portion of their life, and employers had to compensate employees for this as well. When unions won this battle, employers were obligated to provide health insurance and/or pensions to compensate workers for this time.[13] All these debates could happen largely because the metaphor of self-as-property framed how everyone at the time understood the employer-employee contract. This metaphor grounded the struggles over fair employment practices in the United States until the 1980s.

Self as Business

What happens when people start to imagine each individual as a business? When I was doing fieldwork with job seekers and employers, people would tell me that this transformation has deeply affected what it means to work, what it means to have a career, and what it means to be a good employer. While they might not have talked about it explicitly as a shift in metaphor—that is my analysis—many people talked about this change in terms of company loyalty. Now that companies no longer offer security in exchange for the worker's freedom (to switch from job to job, from company to company, or between careers), workers and employers have had to develop new strategies. A career trajectory is no longer based on finding a company to which you can devote your working life. Instead, if you are a worker in the knowledge economy, it means switching jobs regularly as you try to craft a resume dotted with stints at the "right" companies. If the emotion you should feel for a company is no longer loyalty, then what is it? In chapter 7, I argue that you are supposed to experience passion for your work—passion, not loyalty, is supposed to be the new driving emotional force. In addition, as Melissa Gregg documents in her ethnography of Australian middle-class work-life balance, the boundary between company time and personal time no longer holds; people are expected to always be available and willing to respond to workplace demands.[14] After all, a good business is always accommodating and able to provide a solution to a business partner's problem. While owning your self is still at the core of the employment contract, the switch in types of ownership from property to business has involved many fundamental transformations in the nature of work itself.

Owning your self as though you are a business is different than owning your self as though you are property in several important ways. When you are a business, you see yourself as a bundle of skills, assets, qualities, experiences, and relationships. This is a change from how you might see yourself metaphorically as property—as a bounded unit of capacities—although there are parallels. Let me provide a concrete example of the different effects these metaphors have on how you represent yourself as a competent potential employee by comparing the styles of resumes under each metaphor. If you are metaphorically rentable property, your resume is a straightforward list of your previous experiences and capacities. You list where else you have worked, and you list what you are capable of doing on the job. If there is a strategy for crafting your resume, it is a strategy of omission. You might neglect to mention certain experiences that are not relevant for the jobs you are applying for. But even that largely happens because you don't want the resume to be too long. Your resume is a historical record of your abilities; the document does not need to be tailored for every job for which you apply.

Resumes function differently now. They are no longer a historical record; they are now a marketing document.[15] As a result, you are supposed to present the business solutions you have implemented in the past. Under every job listed, the applicant should describe the ways he or she solved problems with as much quantified detail as possible to tout the results. Rather than listing the abilities you demonstrated at a previous job in a limited set of keywords, you provide a short paragraph that describes the business results finessed at different jobs that might be applicable to the particular job you are applying for, especially if those business results can be described using some form of metrics.[16] Tailoring your resume for every job is key, as career counselors will tell people repeatedly.

An earlier resume might describe a job in the following way:

October 1988 to July 1989—Computer Operator, HOME ANALYST of
    PUERTO RICO, Inc., San Juan, P.R.
Duties: Maintained computer input, output, and software utilities,
    updated and organized clients' files, general secretarial work.[17]

Nowadays, a similar job would be described in the following manner:

ACME INC.—Sometown, WY—Midsize marketing and PR firm
Office Assistant, 2006 to Present

Provide administrative and executive support within busy office. Manage executive team's calendar; plan client meetings; prepare reports, spreadsheets and presentations; manage records; and administer database. Results:

*Earned "outstanding" ratings on annual reviews for the past three
  years.
*Recognized for high-quality work, organizational strengths and
  exceptional customer service delivery.
*Praised by supervisor for excellent performance as interim office
  manager (supervising three staff) during her eight-week leave.
*Became Acme's primary creator of PowerPoint presentations and the
  main troubleshooter of MS Office issues.
*Excelled within deadline-intensive environment, ensuring the accurate and on-time completion of all projects.[18]

Every entry in the contemporary resume gives metrics by which an expert reader can gauge previous performances. Yes, each description strings together a list of the duties associated with the job, but for different reasons. The more recent resume does this partially in an attempt to find the keywords that will allow the resume to pass through the computerized applicant-tracking system's screening. In contrast with the older version, the newer version also has four bullet points that provide evidence of what the applicant accomplished for a previous company, statements about how the applicant was evaluated by his or her supervisor, and an assertion of efficiency and competence. It is no accident that these bullet points are listed under the heading of results—resumes today are supposed to present the person as a source of business solutions. This is supposed to allow the hiring manager and recruiter, as the representatives of the company that is hiring, to evaluate whether the potential hire is in fact a businesslike self with which the company would want to ally itself.

When a resume is used under the metaphor that the self is a business, in a sense the resume is part of a negotiation in which two businesses are considering entering into an alliance. The employee-employer relationship is no longer modeled after a property contract. The resume represents the self as a bundle of previously successful business solutions, but not just any random collection of business solutions. Because each business that is hiring presumably has a context-specific set of problems, an applicant is supposed to anticipate the specific problems that a business might be facing. For this reason, there is strong pressure to describe past employment history differently in each job application and to have a tailored resume. When the self is property, a resume lists concisely all the tasks you performed at your different jobs, and you often worry about the right balance between being concise and being descriptive. By contrast, when the self is a business, you struggle to figure out how to present your accomplishments in quantifiably measurable terms, and, at the same time, you try to disguise the teamwork leading to these accomplishments.

To be a bundle of skills, qualities, experiences, assets, and alliances changes which aspects of a person will be valued. For example, what it means to have a skill has changed. Bonnie Urciuoli, a linguistic anthropologist, points out that while *skill* used to refer to manual or mechanical knowledge (perhaps sewing or plumbing), now almost anything that can be tested or ranked can be considered a skill. She focuses in particular on how communication, an activity people simply do as part of daily life, has become a workplace skill or a range of skills that experts can teach and that some people can do measurably better than others. While fifty years ago, no one would have thought that you could get a certificate or even a grade for listening to unhappy employees or talking calmly and patiently to a sick person, nowadays people take it for granted. Now you can measure how well someone listens at a meeting, or speaks French, or runs Photoshop, even if these are all very different acts of communication.[19] Abilities that don't seem linked to each other are all lumped under the category of skills as something that people possess, and the degree to which they possess these skills can be objectively evaluated and compared with other people's abilities. These skills are also always something you can consciously set about improving.[20]

Job seekers face another immediate challenge when they strive to present themselves as a bundle of skills, assets, experiences, qualities, and relationships: that is, how to present their previous history as unified and coherent. Ideally, people's skills are cleanly and easily connected to their experiences, qualities, and so on. And this coherent package can be effectively and clearly represented in resumes and in job application forms, every document that employers ask for believing that these will help them effectively predict future performance. After all, in some sense this is what you are looking for when you hire someone, regardless of which employment metaphor you are using: a predictable hire who will contribute positively to the workplace. Yet this ideal rarely reflects people's complex experiences. This dilemma confronts people who are attempting career transitions, when their past work histories don't easily reveal how competent they might be in an entirely new job, and recent graduates who don't have previous work experience. But, in small ways, this dilemma challenges anyone trying to move from one job to another, because workplaces themselves are not interchangeable— being competent with one group of people and faced with one set of tasks does not guarantee that the person will be competent with another group of people and tasks. Yet in order to switch jobs, the applicant must present him- or herself as a self-contained and internally consistent bundle. Thus, people are constantly finessing how to transform their varied work histories into a unified and coherent self-presentation.

Finally, when you switch from thinking of employment in terms of the self-as-property metaphor to the self-as-business metaphor, you change the basic ideas about what an employment contract is in the first place. When you think of yourself as some form of property, when you "rent" yourself to an employer, you give up some of your freedom to do whatever you want during the day in exchange for some security. When you are yourself a business, you enter into a business-to-business contract with another business when you are hired. Instead of getting a form of security, what you do is create a partnership that distributes responsibility and risk so that every "business" involved can maintain its own autonomy in the market. Everyone is constantly using a means-ends calculus as they work toward distributing responsibility and risk equitably in the alliance.

Let me give you a concrete example. I was talking to a friend's class about online privacy when a woman raised her hand and out of the blue said: "My boyfriend wants to be a cop. But the police department is insisting that he 'friend' a cop on Facebook so that the cop can search his profile to find out if any of his friends are posting about illegal activities." I immediately thought: "That is trading much too much of your freedom for a job. This is a violation of the job seeker's civil rights, *and* his friends' civil rights." I asked the students what they thought of this, that is, how they would react if an employer wanted to use their social media presence as a basis for screening their job application. The class discussion took an unexpected turn, or, at least, unexpected for those thinking of the self as property. A student declared: "This makes perfect sense to me. I want to work as a consultant. And since a company has an image, and I have an image, they will want to check and make sure that my image is compatible with their image."

I did not share with these students the same understanding of what it means to be an employee, and what the employer-employee contract is all about. These students seemed to believe that they need to treat themselves as though they are businesses, and that being hired is entering into a business-to-business contract. Indeed, they are not only businesses, but businesses with brands, just like the company. The students argued that in the knowledge economy the company as a whole and the employee are equally responsible for ensuring that each brand, each public image, does not harm the other. They thought that the company is supposed to maintain a reputation that will not tarnish the employee when the company only becomes a historical marker, a line on the employee's resume—a take on this metaphor that may not be as widespread among business owners as those students would hope. At the same time, there is the far more widespread belief that as long as the employee is working for a company, he or she has a responsibility not to tarnish the company's image in his or her public presentation (on Facebook, Twitter, Instagram, and so on). Upon being hired, an employee's personal brand is understood to be an extension of the company's personal brand, and so when the student started talking about aligning her personal brand with the company's brand, this was an attempt to ensure that neither company nor employee placed an undue burden of risk on the other partner.

This metaphor of self-as-business leads to different kinds of legal issues than the metaphor of self-as-property inspired. Uber, AirBnB, and similar companies are built explicitly on this new model of self-as-business that I am describing, and the legal problems that they face reflect this. These companies allow people to bring their assets (such as their cars or their apartments) and their skills (such as driving and hosting) to the marketplace. From the companies' point of view, they are platforms that enable people to operate as mini-businesses, providing the technological infrastructure that allows people to contract with others for temporary use of their assets (cars, apartments) and skills (driving, hosting). The people providing the assets and services are not employees, or at least that is how the companies see it. Because Uber doesn't own the cars being rented and AirBnB the rooms, they insist that they are only brokering the arrangements and thus are not responsible for regulating drivers' or hosts' actions. But there is an open question at the moment about whether courts will agree. Various lawsuits to determine the status of people who work as freelancers for Uber, AirBnB, and similar companies are moving through the courts as I write. While the legal language involved in these cases revolves around whether someone is a contractor or an employee, for the individuals involved this is also a question of which metaphor people should use to frame the employment contract. Being a contractor or a freelancer is in legal terms the clearest moment in which an individual is a business.

The number of people in the United States working as contractors keeps increasing. In 2015, 17.8 million people worked fifteen hours or more as freelancers in some form, an increase of 33.8 percent from just four years before, in 2011, when 13.3 million freelancers were working.[21] In short, the number of employment contracts which are based on business-to-business contracts, instead of more traditional employment arrangements, has been increasing steadily.

There is another change in the legal landscape of employment these days: job applicants are increasingly seen as potential direct competitors for the companies who might hire them. This is a logical corollary if you start thinking of people as businesses. As a result, new categories of employees are increasingly being asked to sign noncompete clauses before a company is willing to hire them. Law professor Orly Lobel points out, in her book *Talent Wants to Be Free*, that these noncompete clauses are no longer being reserved for a special class of workers, that is, those most likely to know the recipe of the secret sauce behind a company's products. In addition to computer programmers and industry designers, hairstylists, camp counselors, and yoga instructors are being asked to sign these clauses.[22] Now that anyone can be viewed as a business, anyone can also be subject to the legal restrictions regulating business-to-business interactions when they are hired.

What I am describing, the self as a business, is only a model. As a model, it encourages people to analyze social interactions in particular ways and inspires them to put the principles I have been describing into practice. Social theorists know all too well that reality always exceeds and thwarts their models. Because models are reductive by their very nature, social dilemmas are inevitably going to emerge when you put them into practice. And because these are lived dilemmas, writing about them tends to misrepresent how people experience them in one important way. Jean Lave, a fellow anthropologist, has pointed out that writing about these dilemmas can give the reader (and the writer) the false impression that these are problems that can be solved, rather than dilemmas that are always going to be present when you apply a particular model.[23] These dilemmas are not only always present, but in moment after moment, people will continue to wrestle with them anew, choosing to go in one direction in one instance and another, perhaps contradictory direction, in the next. For example, when you imagine yourself as a business, sometimes you want to emphasize how flexible you are, but sometimes you want to show that you have had a lot of concrete experience doing a particular set of tasks. In applying for a job, you are always choosing between seeming adaptable and seeming like you have a predictable skill set. People won't always choose flexibility over predictability. Instead, each time they have to make the choice, they will decide anew which tack to take.

People face new dilemmas under the metaphor of self-as-business. For example, if the self is a bundle of skills, experiences, assets, qualities, and relationships, in practice these aspects of the bundle will be valued differently by different people. Some hiring managers will value skills over qualities or relationships, while others will value relationships over skills. And valuing one element too much over all the others will cause tension. What does this mean when people are hiring? Hiring managers may value hiring someone who is already within their social circle, or is known in their professional networks, over evidence that the person has the skills required. This turned out to be a complaint I heard often from job holders, not job seekers, intriguingly enough. People were frustrated that their work lives were filled with colleagues who they felt weren't good enough at their jobs but knew the right people. From their perspective, networking populated their work interactions with people who were effective at transforming social relationships into instrumental relationships. This might mean that these people were not necessarily good at their jobs, although not always. After all, some jobs depend entirely on being able to turn your friends and acquaintances into business opportunities, but not all jobs. Current employees who were frustrated by such hiring decisions also felt that networking ultimately undercut efforts to create the best workplaces with the most competent employees because people often have to make calculated choices about whether they are going to enhance their networks or enhance their skills. There is only so much time in every day, and you sometimes have to choose between meeting people or practicing a skill. The people who choose repeatedly to enhance their networks might survive the transitional moments between jobs, but that can come at the cost of enhancing their skill sets. Here is a moment in which what makes someone good at *getting* a job is different than what makes someone good *at* a job. In the hiring process, because this bundle that makes up the self is in fact complex and often uneven, valuing one aspect—be it skills, experiences, qualities, or relationships—over all else could put the selection process for job applicants and the actual workplace practices at odds with each other. This is only one of the lived dilemmas that the self-as-business metaphor produces for people. There are countless others, and in the rest of the book, I'll talk about some of the most common ones.

How I Studied Hiring in Corporate America

I could have studied corporate hiring anywhere in the United States, although when I was wondering where to go, I thought urban areas would probably have more people looking for jobs and more nonprofit organizations geared toward helping the unemployed. In the end, I didn't choose a location in advance. I applied for grants from a number of different organizations and let the funding bodies' decisions determine where I would do this research. I was lucky—in 2013–2014, Stanford's Center for Advanced Study in the Behavioral Sciences offered me a fellowship. I spent the year based at Stanford, traveling throughout the Bay Area talking to anyone I could find who was willing to discuss hiring with me. I interviewed everyone involved in the hiring process, from job seekers, and the career counselors who advise them, to all the people evaluating job applicants—hiring managers, recruiters, and HR managers.[24] I also talked to people who had recently gotten jobs and people who had recently quit jobs. When I found myself too swamped with interviews with knowledge economy workers, I hired Lori Hall-Aruajo to supplement my too few interviews with blue-collar workers. In all, Lori and I conducted 165 semistructured interviews in which we discussed hiring and job-seeking strategies. This total includes my interviews with graduating students and career counselors at Indiana University before I started the Bay Area research.

For the most part, I thought of myself as being in the Bay Area, not in Silicon Valley. I didn't want to focus on the tech industry only. I also wanted to know how recruiters hired temporary caregivers and how lawyers looked for work. But, of course, because I was in the Bay Area, I did talk to many people involved in the tech industry. For chapter 5, I took advantage of the fact that I was doing fieldwork around so many new start-up companies to research the people who are developing new technologies to address perceived problems in the hiring process.

In addition, I attended 54 workshops in San Francisco and the surrounding area that taught job seekers how to use LinkedIn and other social media, as well as how to write resumes, network, interview, and develop their personal brands. A number of remarkable organizations in the Bay Area had, as their primary purpose, helping job seekers, any job seeker, find work. Some of these organizations were national nonprofit organizations, some were the brainchild of a single man or a group of churchgoers, and others were county or city government organizations. They regularly held free workshops for anyone who wanted tips on how to write a resume or fashion a LinkedIn profile. I also conducted a series of focus groups at community-based organizations geared toward helping professionals find jobs.

After all my interviews and observations, I was still left with a number of questions about whether the advice circulating in the workshops I attended is as effective as people claim. My interviews weren't helping me get these answers, in part because I was talking to job seekers who could tell me how they got an interview, but many of whom were still hoping to get the job. And those employing applicants often didn't seem to notice the techniques that job applicants were supposedly using. No one on the hiring side, for example, talked to me about ever noticing an applicant's personal brand. Since interviews and the participant observation I could do weren't supplying my answers, I decided to take another route. An organization in the Bay Area had weekly meetings in the local city council chambers during which every member who had gotten a job that week came to the front of the room to tell his or her success story. People were allotted three minutes to tell how they got their job offer and to thank everyone who had helped them along the way. The stories often took longer than three minutes, but there was always someone in the front row with a timer, warning the speakers to be brief and concise. These success stories had been recorded since January 2012. The organization generously provided me with access to all the success stories that had been recorded between then and May 2014. For three long weeks, slightly dazed and turning to chocolate to get through this task, I watched each and every story, creating a database of 380 success stories. This database, supplemented by my interviews, is my source of data when I mention any statistical analysis of the relative effectiveness of different job-search techniques used by the population I am studying.

The most interesting finding I have from compiling this database is that Mark Granovetter's marvelous study of how people switched jobs in the early 1970s, *Getting a Job*, should be revisited because of the dramatic change in everyone's media ecologies since the 1970s. Granovetter found that 56 percent of people heard about jobs through weak ties, that is, that weak ties were a crucial source of information for finding a new job. In chapter 3, I argue that new technologies have changed the kinds of problems people face when searching for a job. The problem is no longer, as it was in the 1970s, discovering that the job opening exists in the first place. Instead, job seekers' major problem is ensuring that someone notices their resume now that so many people are applying to every job opening. When you want your resume to be noticed, it turns out that workplace ties—people who can speak to what you are like as a worker—help white-collar job seekers much more than weak ties do (61 percent of my sample were helped by workplace ties, and 17 percent were helped by weak ties). This is not to say that Granovetter's study is wrong, but rather that it is a grounded snapshot of a historical moment. New technologies such as the internet have affected how information about jobs circulates and have transformed *when* in the hiring process this circulation of information poses a problem for job seekers.

Hiring, Disrupted

In the rest of this book, I explore what happens when people understand hiring as a business-to-business contract. Seeing hiring as a business-to-business relationship in the contemporary United States has meant thinking about hiring as a short-term contract with an end (however indefinite). While people are at a job, they are supposed to always be planning for the next job, and if you have hired someone, you should always be anticipating having to replace that person. This affects how you build relationships in a company and how the company circulates information. In the abstract, business alliances need not be short term, but in today's America, there are fewer and fewer pressures or structural reasons for business relationships to last. By contrast, there are many reasons for business alliances to be relatively short term, especially given how much US companies rely on being "disruptive" or offering the latest new or improved product to justify their value.

When hiring becomes a short-term contract, people have to figure out how to be ethical in such short-term relations. In my interviews, people kept talking about what the obligations of employers and workers might be, especially now that everyone has to assume short-term employment. And for my part, I kept wondering about what is in place to ensure that employers and employees agree on and meet their obligations. When you aren't planning to stay at a workplace for very long, what keeps you treating people at that workplace well, whether you are someone's boss or someone's coworker? Why would you behave in ways that improve someone else's life if you aren't going to be interacting with them after a year or two? As I note about networking in chapter 3 and quitting in chapter 7, people would claim that they behaved well toward everyone they worked with in part because of this uncertainty: you never know when in the future someone else will be just the person you want to hire, or in a position to hire you or help you get hired. But is that enough to keep people behaving well?

Both employers and employees often found the idea of having a reputation and needing to protect it to be a good enough incentive to behave decently. Getting a bad reputation not only risks your relationship with that person or corporation, but it also risks your ability to create successful future business relationships. This emphasis on reputation, while not new by any means, requires different strategies now that information circulates on the web. Often this information circulates in relatively short snippets of text. As a result, nowadays it is probably safer for a company intending to do something harmful to its employees to figure out complex ways of doing so that avoid having clearly responsible actors and have a lot of moving parts. A company brand is less likely to be easily damaged by strategies that can not be described quickly and succinctly online.

The problem that people in my fieldwork ran into time and time again, whether they were hiring or being hired, is that people's ability to present themselves as competent job applicants can have little to do with what they will actually be like once they have the job. All too often, the categories and practices used for beginning an alliance are only tangentially related to the practices necessary for maintaining a satisfactory one. That is, the knowledge someone brings to being competent at writing a resume or interviewing for a job may have little to do with what that person needs to know in the workplace. And as I mentioned earlier, the networking a person does in order to get a job is often just a subset of the kinds of social skills needed to be good at that job. This is an old problem, and certainly not a result of this new metaphor. But it does, however, become a much more common problem when business-to-business relationships become the model for all other forms of work relationships.

When the self-as-business metaphor replaced the self-as-property metaphor as a way to understand the employment contract, the same logic began to encourage businesses to be as flexible and dynamic as possible and to see long-term stable commitments to a workforce as obstacles to this agility. Since the 1980s in the United States, companies and self-help gurus have been encouraging workers to value agility in place of company loyalty and other established forms of stability by insisting that workers view themselves as businesses in their own right. This new model transforms the employer-employee relationship into a business-to-business contract. The change has been gradual, and while widespread, it is still in the piecemeal process of being adopted and voiced. Not everyone believes in this new metaphor, not everyone agrees with it, and not all the infrastructure and actual practices used during hiring reflect it. However, if you are unemployed for any length of time in the United States, at some point you will be told to view yourself as a business and to act accordingly. And anyone hiring these days comes across job candidates putting this model into practice. In this book, I discuss some of the challenges that emerge when hiring or trying to be hired as a business, that is, the contradictions that only become apparent when you try to put this logic into practice. By doing so, I hope to point out the fault lines, the moments when people who are ambivalent about seeing themselves as businesses might start wishing that there was another way, and may even be inspired to act and think differently.

One	**You Are Just like Coca-Cola**
	Selling Your Self through Personal Branding

A few years ago, I attended a workshop on personal branding with about a dozen undergraduates from Indiana University, who were, no doubt, wondering what I was doing there. The too-large room contained as many pizzas as people. Pepper, a Google recruiter who was leading the workshop, had brought them, clearly expecting a better turnout. But she wasn't intimidated by the empty room. In her midtwenties, dressed in jeans and a sweatshirt, she played music and was dancing around before the workshop began.

After we got started, Pepper asked us to spend thirty seconds introducing ourselves to one of the people sitting next to us. We then had to write down three words that we thought would describe our neighbor. I was a bit mortified that the slightly clueless but determinedly goodwilled undergraduate I chatted with decided to describe me as "sweet." I struggled to describe her. She had said such nondescript things about liking Bloomington and traveling to Boston for winter break. I ended up describing her as "earnest, curious, and tasteful." After this exercise, Pepper explained that the reason she got her job at Google was related to this adjective game, an anecdote that was obviously part of her effort to convince her audience that fashioning a personal brand was now an essential part of searching for a job. In every personal-branding workshop I have attended, the instructor assumes audience members don't know what a personal brand is and are not already convinced that they need one. The first third or first half of every such workshop seems devoted to convincing the audience of the value of branding techniques.

24   Without openly reminding the audience that this was just as the financial crisis was unfolding, Pepper explained that she had interviewed at Google in 2008. After she got hired, she wondered why and asked a man who had been on the hiring team that interviewed her. He told her that he didn't remember the answers she had given during the interview, her GPA, or her resume. What caught his eye was that she was one of the most positive people he had ever met, and he knew that she not only would make his work life better but would be well-suited to a customer service position. "You didn't know this," he went on, "but I had actually just started at Google about three months before I had interviewed you. And I worked for an auto company in Detroit that was failing. The attitude at work was so negative that I had to leave. When I saw you, I knew that I wanted someone like you on my team." Pepper concluded her story: "So I didn't even get three adjectives from this man who had interviewed me. I got one. And that hands down is what got me my job at Google." For Pepper, this story demonstrated the unquestionable value of branding. I see things a bit differently.

Figuring out your personal brand involves deciding on the three or four words that capture your essence, that describe what personal-branding aficionados will call your authentic self. Personal branding as an idea may simply involve marketing yourself, but what exactly does marketing yourself involve? The idea is accompanied by very specific practices nowadays, ones that have evolved out of the strategies American advertisers first developed to brand commodities. Personal branding is different from making sure that you have a good reputation, although people may talk about personal branding and a reputation as interchangeable. With personal branding, and unlike managing your reputation, there are relatively standard techniques which career counselors and workshop leaders advocate that you use to figure out what self you should market, and equally standard practices for promoting your personal brand.

It is telling that both in Pepper's exercise and in her example, people are supposed to sense the words that underlie your personal brand without explicitly being told what these words are. You know that your brand is successful when the qualities you try to exude, encapsulated in the handful of words you have already selected for yourself, are then reflected back to you in the words people choose to describe you. For Pepper, the mark of her branding success was that she and her interviewer agreed that she was a positive person.

But when Pepper's interviewer was explaining why he hired her, he was actually reflecting on a complicated socioeconomic situation: the collapse of the automotive industry in the wake of the recession. Yet the only way it features in this story is as a commentary on the kind of emotional labor her interviewer would like his coworkers to perform. He wants to be surrounded by happy people, not people who are worried that a dire economic situation will lead to massive layoffs. Maybe his coworkers in Detroit would have been positive too if they had secure well-paying jobs. Branding encourages you to focus on a person's supposedly unchanging qualities and to ignore contexts.

Pepper is one of many people in the United States urging job seekers to begin developing a personal brand. There are any number of workshops for job seekers on crafting one's personal brand, and it is a common theme in self-help books and internet articles. At the same time, none of the hiring managers or recruiters whom I spoke to mentioned paying attention to applicants' personal brands. They talked about focusing on people's qualifications and worrying about whether a person could actually do the job for which he or she was hired. No one said "Well, I try to hire people with good personal brands," or described noticing any of the practices applicants are supposed to do to promote their personal brands. In fact, many of the screening mechanisms people use to select job candidates don't give any employer enough time to see most of the ways applicants will try to build personal brands.

As far as I could tell, personal branding was a concept and set of activities that career counselors and motivational speakers promoted, job seekers engaged with, and those hiring ignored. Yet whenever I was critical of personal branding in conversations with job seekers, they would defend the idea. What makes personal branding such a compelling concept nowadays for people who believe that anyone can get a job if they just use the right techniques?

I think it does a couple of things, neither of which is particularly helpful for job seekers. First, it is a logical (but not necessary) extension of how people currently think about the employment contract, a contract that no longer includes the promise of stable long-term employment. Second, it provides a set of prescriptive standardizing techniques for managing something that baffles many job seekers these days—constructing an online presence that signals to others that they are not only employable but desirable as a worker. But these standardizing techniques don't add much useful information for employers.

You Are Just like Coca-Cola

If people now need to manage themselves as though they are businesses, then it isn't that far of a stretch to think that a person can have a marketable image. Indeed, when Tom Peters first popularized the notion of personal branding, in 1997 in a *Fast Company* article titled "The Brand Called You," he insisted that we must now all brand ourselves because "we are CEOs of our own companies: Me Inc." The logic seems to run as follows: if we are all now companies, then we have to start doing what companies do, including marketing ourselves. The possibility that people need a brand on the job market makes intuitive sense for people committed to seeing the hiring relationship as a business-to-business contract.

But once enough people in the United States think that it might be a good idea for every worker to have a brand, then another question immediately arises. How do you go about making one? Most marketers would probably say that this is extremely easy, that their techniques allow absolutely any kind of entity to be branded—sodas, vacation spots, companies, college experiences, countries, and people. What a marketer would do is take techniques developed for associating an object with a brand and then simply substitute a person. But it turns out that the ways in which a person is *not* like a commodity guarantee that people will face a range of relatively pre-dictable problems when they try to brand themselves.[1]

When advertisers develop a brand for an object, they try to endow it with a personality. A group of marketers might sit around a table trying to answer the question: if Diet Coke was a person, how would you describe that person? Once they have a list of adjectives, they then have to figure out how to imply these adjectives to a broad audience without actually uttering them. This supposedly is based on how people reveal their qualities persuasively. You might not be convinced by someone who openly tells you "I am a genius." As branding experts will tell you, all you have discovered is how they think of themselves. Yet if you witness them coming up with brilliant suggestions, you might be inclined to think of them as a genius. As an example, anthropologist Robert Moore describes how one woman teaching branding to her coworkers explained that brands function like a beautiful woman at a party. She does not walk into a room and loudly announce, "Everyone, I just want you to know that I am beautiful." Instead she projects this beauty, and if successful, everyone at the party associates her with this attribute.[2]

As Celia Lury points out in *Brands*, in order for an object to have a personality, the branded object's personality has to be a collection of abstract traits. You have to believe that a personality does not depend on context. That is, you are not supposed to be hardworking only when you are in a workplace where you like your coworkers a lot and you have a task to do that you find fascinating—perhaps figuring out why a car always stalls. You are supposed to be hardworking in every situation. When a brand personality is associated with an object, it is a very abstracted and reduced notion of a personality. For example, Diet Coke is meant to be associated with a set of qualities regardless of the surrounding conditions, such as how it was stored or the context in which someone drinks the beverage. And an object's personality does not have to be based on a real person, but a generic idea of what a personality is.[3] So these words don't apply to how someone is in a specific situation. In practice, the qualities are a fairly limited collage of generic characteristics.

28   Because branding depends on crafting a context-free personality, it can be used in unexpected ways and for unlikely populations. Ramina was a career counselor who explained to me at length why she thought branding was a profoundly influential strategy to teach men and women who had been recently released from jail and were looking for jobs. She had been involved in a project organized by parole officers geared toward helping their parolees navigate complicated and not-so-accepting job markets, a challenge she loved. Ramina thought branding encouraged them to see what value they had to offer others and to begin to appreciate the skills they inherently brought to any job. What was fascinating to me was that she took the techniques that personal branding requires, of separating personality from context, and asked parolees to reimagine their past experiences along these lines.

She told me how she did this. In her workshops, parole officers were lined up along the walls because having so many parolees in one room is considered dangerous. She would begin by asking everyone to talk about their skills. If a woman told Ramina that she was one of the best drug dealers in her neighborhood, Ramina would respond enthusiastically: "Let's take that illegal element out of it and let's focus on your skills. What did it take to make someone trust you over and over to come in and buy drugs from you?" The woman might explain how good she was with customer service—she was honest. If another parolee, say a man, wasn't a former drug dealer but a former drug user, Ramina would ask him how he knew which drug dealer to approach. He might tell her that he was particularly good at reading people, and she would encourage him to note that skill. Ramina asked parolees to distinguish between the conscious strategies they used to navigate a situation and what they were actually doing in that situation, the illegal activities. Because a brand personality is context free, it doesn't matter whether you are good at figuring out who is a trustworthy drug dealer or who is a trustworthy car salesperson. What is important is that you have a talent at determining who is trustworthy.

When people try to brand themselves, they are using techniques designed to associate an object with a personality, techniques that had to be radically simplified in order to be effective precisely because objects don't engage in the world in the complex ways that people do. Diet Cokes aren't moody when talking to an ex-boyfriend and relieved and happy after getting a good performance review at work. Even in Ramina's version, branding oneself is a simplifying exercise in which one ignores all the reasons that one might be using one's particular abilities in the first place. She has to teach her students approaches for ignoring the context in which they were acting and for focusing on a specific way of interpreting their actions— something her students may never have done before. In general, many of the difficulties that people experience in trying to brand themselves come from the fact that they are using techniques developed to associate objects with personalities, and, to be effective, these techniques have to distort what a person's personality is often understood to be.

When Do Objects Need Personalities?

This raises another question: why would objects need human personalities in the first place to appeal to consumers? It turns out that companies started to link objects to personalities when mass-manufactured goods became widespread in the United States. Companies were trying to reassure customers who were more comfortable buying goods from trusted shopkeepers who sold undifferentiated merchandise (for example, a pound of sugar or a bushel of corn). For the most part, these goods were provided by local producers— customers might have gone to church with the farmer who grew the corn that they ate.

Historian Bruce Schulman writes about the rise of product brands, describing how the rise of mass-manufactured goods significantly altered people's shopping experiences from the 1890s onward. He points out that before the 1890s, people purchased goods from sellers whom they knew—the local pharmacist compounding his own medicines, the peddler who came every six months, the local grocer who bought his produce from neighboring farms. They also bought tomato sauce or cough syrup, not Heinz's ketchup or Lydia Pinkham's cough syrup. They knew the tomato sauce would be good because, perhaps, they knew the shopkeeper's wife was following her mother's recipe, and her mother had a local reputation for making the best in town.

In the 1890s, consumers had to learn how to trust new supply chains and new producers. Shoppers began to encounter goods that were differentiated because they were associated with a particular company, not a local shopkeeper or local farmer. People increasingly had access to large department stores, and even local dry goods stores began to stock a range of mass-manufactured items. These items were packaged, which often hid the product from view and touch. Wider distribution, longer supply chains, and packaging all combined to create a quandary for consumers—how could they know that the goods they were buying were of good quality?

Companies decided to mimic the personal relationships consumers had developed with local shopkeepers, and created recognizable and vivid figures to accompany the products. Schulman writes: "National advertising campaigns testified to the purity of products in a market where fears of adulterated goods ran rampant. Recognizable trademarks and packages became old friends, easing the transition to a brave new world of commerce. Some manufacturers even created characters, asking customers to write with questions, recipes, and problems."[4] In the early twentieth century, providing companies with a brand personality was literal in a different sense than it is today. The companies were not ascribing a set of abstract personal qualities to an object. Instead, they were providing a corporate character, early precursors to characters such as Aunt Jemima or Mr. Clean, to take the symbolic place of the local producer or shopkeeper and thus reassure consumers. Brand personalities alluded to the complex histories and familiarity that consumers had with individual sellers, encouraging them to create similar ties and thus similar forms of loyalty to a company producing a broad range of mass-manufactured products.

Nowadays, the idea of buying mass-manufactured goods is much more commonplace, and consumers no longer have the same concerns about how to find a company and its products trustworthy. Instead, consumers may have a different set of concerns about supply chains and companies' reliability, especially if they are interested in fair labor practices or environmental pollution. Yet people still encounter the vestiges of these older corporate attempts at reassuring customers through the ways that brands today are intertwined with personalities.

Over the course of the twentieth century, brands were increasingly used to differentiate products. The purpose of branding shifted away from primarily inspiring a consumer to trust a company. Instead, as Robert Moore points out, brands became a solution for a dilemma that companies have whenever they are selling a product: how does a company convince potential consumers that, for example, two Starbucks lattes taste the same, especially if they are made by baristas in different stores? One solution is trademarks, symbols designed to suggest that the designated objects and events (say, ClubMed vacations) provide interchangeable and predictable experiences.[5] Pepper mentioned this concept in her workshop on personal branding when she asked the audience to describe why Diet Coke has such an effective brand. One male undergraduate said: "I like the way it is always the same." Pepper promptly agreed: "You can go anywhere, and get a Coca-Cola. In Italy, when I was downing Diet Coke, when I was studying abroad, it was called Coca-zero. But it still tasted pretty close to the same thing, which is definitely something which makes them unique." In this workshop, Diet Coke's success as a brand was directly discussed in terms of its predictability—you can find it anywhere you travel, and you always know the taste you are purchasing.

Your Inner Manager

Personal branding developed to solve a market-specific problem for job seekers, one that has resonances with how brands are supposed to make objects seem predictable. Tom Peters argued in his 1997 article "The Brand Called You" that people need to brand themselves because career trajectories are no longer clear-cut vertical paths. He writes: "A career is now a checkerboard. Or even a maze. It's full of moves that go sideways, forward, slide on the diagonal, even go backward when that makes sense. (It often does.)" Once staying with a single company is no longer an ideal career trajectory, people have to figure out how to craft a coherent narrative about their work history for recruiters and future hiring managers. People are switching jobs so often, and changing what they do in each job, that the already established quick shorthands developed to explain who and what someone is as a worker no longer are reliable. What Peters and other career counselors argue is that branding yourself is an effective strategy for explaining easily why a work history that may look like a maze or a checkerboard at first glance in fact has an internal logic of its own. Just as branding is supposed to help convince consumers that one Starbucks latte will be like another Starbucks latte, regardless of whether it is made in New Orleans or Paris, so too with personal branding, which is supposed to convince employers that they are hiring a stable, predictable person, regardless of the varied complexities of where that person worked previously. Or as Pepper put it: "You are just like Coca-Cola, you are super unique, and there are plenty of things that make you stand out, and we just have to figure out a sick marketing campaign to get you there."

In one of the free workshops for job seekers that I attended at the San Francisco Public Library, Nick explained that as you manage your personal brand online, it helps to link this brand to one of your hobbies. If you build your brand not only around qualities but around culturally acceptable hobbies—biking, say, or hiking—you can fashion an image of yourself that will create a sense of coherence no matter how many career transitions you might make. In responding to someone's question about the difficulty of rebranding your self while trying to make a career transition, Nick explained that if your brand only revolves around your professional practices, it might become difficult to switch to a new job: "It is really important not to rely too heavily on just your professional experience when you are building your personal brand, right. That way it becomes easier to make a shift because you are only changing a few small things, your core interests will stay the same." For Nick, the stability promised by maintaining a consistent and coherent image constructed around one's leisure activities serves as a useful counterpart to the constant flexibility required by contemporary employment conditions. Of course, this presumes that you keep your hobbies and outside interests constant. Something has to be kept stable when you are imagining yourself to have a brand, and people have different solutions for where to locate that stability.

If brands are, in part, ways to communicate to consumers that they will have the same predictable experience with a product every time they use it, then this becomes a much more complicated task when applied to a person. From an American perspective, people are not expected to be stable and consistent in the same way that objects are supposed to be. For instance, people's behavior changes depending on context in a way that objects don't.

It can be difficult to frame yourself as a set number of qualities that mark you as distinctive and yet still a good worker. Justin, who taught career development courses at Indiana University, explained to me that he had a particular class exercise for teaching students how to fashion their own brands, similar to Pepper's exercise. He would ask each class member to write three words or phrases that describe his or her individual essence. These words have to be specific enough that they mark what makes a person unique yet still well within an appropriate rubric. When I asked Justin what types of words and phrases did not work and why, he explained that *diva* or *liking the outdoors* were not good choices because *diva* is, in a sense, too distinctive while *liking the outdoors* is not distinctive enough.

In workshops, people aren't given a list of words and asked to choose the keywords that will be the basis for their personal brands. This would undermine the ways these words are supposed to come from within, reflecting the unique personality of that person. Yet I had the vague sense that there was a widely understood set of words people turned to, since certain words kept cropping up: *passionate, reliable, hardworking, compassionate, committed, dependable, flexible, creative, resourceful, enthusiastic,* and so on. Each person might choose his or her own set of three or four keywords out of this broader set, but no one chooses unexpected terms, perhaps negative words (*morose, irritable, melancholic*) or ambiguous words (*sarcastic, skeptical*).

The next step is figuring out how to demonstrate these keywords. One master's student I interviewed explained how he carefully chose the words *simplicity* and *elegance* to be the markers of his professional persona, uniting all aspects of his web presence. He showed me his website—it was black and white. He thought a monochromatic design with subtle accents of color would lead viewers to think "elegant." All text was formatted in lowercase, even his use of the pronoun *I.* For him, lowercase helped signal simplicity. These were visual and typographic strategies he was using to imply qualities without openly stating them, just as a beautiful woman is supposed to do, according to branding experts.

Both of these interviewees described strategies for fashioning a self as a brand personality that is flexible enough to appeal to employers but stable and distinctive enough to be recognizable and coherent. This in itself, however, can be a dilemma. When the self is a business, you are meant to be constantly enhancing your skills, alliances, and assets—you are a self constantly in the process of transforming into a better version. How, in this process, does this self remain stable enough to be recognizable and yet flexible enough to be hirable?

This constant enhancement is produced by the kind of reflexivity that thinking of yourself as a business encourages: the self as a conscious manager. When someone decides that their essence is reducible to words, such as *simplicity* and *elegance,* what they are outlining is the ways in which their own collection of skills, experiences, assets, and alliances is to be managed. A person is both signaling the set of choices he or she is making and the logic or style with which these choices are made. In general, the personal brand is supposed to be a standardized form that intertwines flexibility and coherence. By making visible the supposedly fundamental characteristics of the reflexive manager, the personal brand reveals precisely what could appear to be a contradiction: how can someone always be willing and able to transform and yet remain a cohesive self?

Unique and Authentic in the Right Way

One way that I have seen career counselors openly get around this conundrum when they are teaching a workshop on branding is to stress that everyone has an authentic self, and the brand one develops is supposed to reflect this authentic self. This authentic self supposedly doesn't change in response to experiences, doesn't have epiphanies, conversion experiences, or any other form of life-changing experiences. Personal branding allows people to represent themselves as both flexible and coherent, that is, able to move fluidly and effectively through multiple contexts, yet with a managing impulse that makes them appear predictable to potential employers. It is this managing impulse that defines what constitutes the authentic self.

Intriguingly enough, no one talked to me about their authentic self while I was doing fieldwork except in the context of branding. Perhaps Americans only worry about displaying an authentic self when they are faced with the dilemma of marketing themselves as if they were Diet Coke. After all, from an American perspective, objects are not the same kind of mix of stability and instability that people are in practice. Americans ascribe a whole different set of explanations for describing why an object might behave in ways that they didn't expect than the set they use to explain why a person has behaved in ways that they didn't expect. But the notion of an authentic self helps make the comparison between person and object possible and the use of these techniques seem reasonable.

When workshop leaders talked about how important it was that one's branded self was properly aligned with one's authentic self, their explanations often had moralistic overtones. Some talked about how a branded self that wasn't anchored in an authentic self would be ineffective. Workshop leaders liked using the following quote in their PowerPoint presentations, attributed to personal-branding advocate Dan Shwabel: "Be the real you because everyone else is taken and replicas don't sell for as much." At some point during the workshops I attended, every speaker stressed that faking it would be wrong. Pepper managed this with her usual panache: "I know that I have an awesome personal brand, and I know that everybody wants it, just kidding. But it is never going to work, if you walk out of this saying, 'Oh gosh, Pepper had a great brand, that's who I am going to be.' That's not authentic, that's not who you are, and that's not what you should be doing with your personal brand."

A moral way of being was at stake, a point which became clear when Pepper brought up the example of Lance Armstrong. "LIVESTRONG on those bracelets, oh my gosh, those are the best," Pepper gushed. "So he even had like a branding campaign around his personal brand which said 'strong' in the actual name of the brand. He was strong, he was fast, he was powerful, he was successful. He won the Tour de France and beat cancer in the same year. So what happened?" For Pepper, Lance Armstrong was driven to doping because of the incongruity of the brand he had developed for himself and his actual capacities. He was forced to turn to illegal means to ensure that he could in fact live up to his branded image, a trap he would not have fallen into if his brand had been more accurate in the first place. Her moral parable ended with a caution: "So make sure as you are thinking about this uniqueness, this personal brand, make sure it is super-authentic to who you are."

Not everyone I spoke to wanted to make sure that their branded self was aligned with their authentic self. Judy, a woman in her fifties talked about personal brands in a way that reminded me of earlier understandings of what it means to be a good worker, that is,

moments in which you have to perform emotions that you don't
necessarily feel because that is the job.[6] She told me that she liked
the idea of personal branding a lot, because it allowed her to create a
work persona online that served as a shield for her authentic self. By
carefully crafting a safe professional persona, she was able to ensure
that potential employers did not know her private self at all.

I told her that it sounded to me as if creating a personal brand was
the work a person has to do as a waitress, just on a larger scale. She
agreed, and admitted, however, that whenever she thought about
creating a personal brand, it seemed like too much work. She wasn't
sure she wanted to put in the time to create one. I was fascinated
that she thought of branding as offering protection from the pry-
ing eyes of employers, since her current job was as a cashier for a
large chain store. She was constantly being monitored and being told
that she had to create a positive experience for customers. This is a
kind of emotional work that makes you disconnect from whatever
you are actually feeling.[7] It made sense to me that she would want
branding to offer a similar protection. Judy wasn't convinced that her
brand had anything to do with her "authentic self." Instead she rein-
terpreted personal branding according to older, and class-specific,
understandings of how people's work lives are supposed to be separate from their personal lives.

Judy was an exception. Most people I spoke to were very concerned with aligning their branded self with their authentic self. Yet
linking your brand to your authentic self can raise a whole set of
dilemmas for people because it means that you have to know who
your authentic self is.

Dorothy, a woman in her sixties, had been a career counselor, had
transitioned to human resources, and was currently looking for a job.
When I met her at a workshop, she was struggling to figure out what
her brand was. Later, over a latte at Starbucks, I admitted to her how
bizarre I thought it was that the branded self should have anything to
do with the authentic self, that I didn't think her difficulty finding her
brand had anything to do with an identity crisis. Dorothy disagreed:

Well, it's more a search for the authentic self, I guess. I think I'm probably a lot more in touch with my authentic self than I think. But my opinion is I need to be more in touch. So that's what the rebranding is about. I've used some of the key words to describe myself, draw people in and it's not like I've been lying or anything but it's like, what do I really want to emphasize in terms of my skills? . . . And so I haven't been able to update my LinkedIn profile. I haven't been able to rebrand it.

Dorothy was stymied; she couldn't begin the work of branding until she figured out more about her authentic self. Perhaps it was an issue of what part of her authentic self she wanted to highlight, which implies that not all authentic selves can be easily reduced to three or four words. Or perhaps she just sought to be clearer about who she really was, so that she could more effectively align her authentic self with her branded self.

Pepper too warned that one of the hardest parts of branding was the self-reflection it required. She recommended beginning by answering a handful of questions, such as "What is your super-power?" and "What do you do better than anyone else?" These questions, however, do not easily lead to the three or four qualities one should claim capture one's authentic essence. Instead, they are trying to uncover a sense of uniqueness, but a very particular version of uniqueness that emerges only out of comparison with other potential workers.

The kind of uniqueness that informs personal branding is not entirely new. In fact, cultural critic Walter Benn Michaels argues that it became a dominant way of thinking about what it means to be an individual in the United States between the 1860s and 1880s, in the wake of the Civil War. At that time, people first began to understand themselves as "individuals individualized by their place within the system"[8] With this phrase, Michaels points to a particular dynamic, one in which you become an individual or distinctive person because you are already within a system that compares and contrasts you in some way to everyone else. You can't be a unique mixture of boldness, reliability, and creativity without comparing yourself to everyone else in terms of a predefined set of qualities that they might have. What you use to show that you are unlike everyone else—the signs you use to make sure that your individuality is recognizable and visible—is, however, already predetermined by various institutions that rely heavily on standardizing technologies.

Michaels's discussion of Civil War uniforms is one of his more captivating examples of how this concept of individuality came to be widespread. He argues that clothing sizes did not exist until the Civil War, when, for the first time, armies required mass-produced uniforms (and the requisite technology was available). This was the moment that clothing sizes, such as small, medium, or large, became widely available, and at the same time, a person learned the ways in which his or her body shape could not so easily fit into these standard categories. A person's chest or arms became distinctive inasmuch as they did not match the way the rest of the body conformed to a medium or a large size. Maybe your chest was average but your arms were long. At this period, uniqueness or individuality emerged as the particular pattern by which you assembled or combined standardized forms for indicating that aspects of your self are distinctive within an already established system.[9] Figuring out how you are distinctive when it is a clothing system isn't that difficult—you can figure out pretty easily that most of you will fit into a size 8, but you always have to roll up your pants because your legs are a bit too short. It can be much more complicated to figure out what makes you a unique worker, especially when so much of actual work is about getting a task accomplished in a competent, standardized way.

Personal branding pivots on this type of uniqueness, in part because it is so crucial for how objects acquire brands as well. As communication scholars Daniel J. Lair, Katie Sullivan, and George Cheney explain: "Brand products were marketed as *unique* goods able to provide *unique* advantages to consumers; it was the brand name that distinguished a product—for example, Spic'N'Span— from other household cleaners."[10] Yet taking the uniqueness ascribed to an object or product line and using it to frame a person can be a complicated task.

In a workshop I attended on the elevator pitch, or how to describe yourself in thirty seconds, Lucy had a heartfelt outburst about how frustrating she found it to express her uniqueness. "I think what I do is perfectly normal," she exclaimed, "and I don't see how I am any different from all the other people that I have met at Promatch, JVS, and networking groups. I don't see myself as particularly unique, so I don't know what to do!"[11]

I definitely felt for Saul, the workshop teacher, when he tried to reassure Lucy that she would be able to easily figure out what made her unique if she just asked other people to help her identify her strengths. Lucy was such a vivid character, in the way that no two crazy cat ladies are alike. She was memorably awkward, running after anybody who started to leave the workshop early and insisting on exchanging business cards with them then and there. They tended to look pained when she did this, and one man absolutely refused to acquiesce to this perfunctory business card exchange. She took to such an extreme the suggestions that she must have received at other networking workshops that she made networking seem like parody, and yet she was completely serious. Lucy, in short, was the kind of person who someone might say diplomatically about her that she was unique, and others would nod straight-faced, understanding in that moment that the term *unique* meant a flair for imaginatively getting social norms wrong.

Saul was faced with both reassuring Lucy that she was indeed unique and transforming the potentially pejorative implications of how clearly distinctive she was into a perspective that might help her get a job. When I suggest that there are many forms of uniqueness, and personal branding depends on a very specific kind, Lucy's easily noticed uniqueness is a good example of the wrong kind of uniqueness for personal branding.[12]

People can also decide that they are particularly talented at something that isn't generally appreciated as a reason to hire someone. One job seeker in her late fifties whom I spoke to explained that she was remarkably good at understanding the potential pitfalls of plans people suggested at work. Yet she was an administrative assistant and wasn't in a hierarchical position where this analytical approach was going to be valued. And, like many other office workers I talked to, she felt that her managers only valued people who enthusiastically supported proposed plans. So she had to figure out something else she was good at, and she hit upon her gift at organizing office birthday parties. "I am really good at kind of seeing people," she told me, "at being supportive and doing things that will make people smile." When she found out that one of her managers liked comic books as a kid, she said she made a mental note: "So for his birthday, I went and found comic books from the year he was born. I put his face onto one of the superheroes, and I turned it into a little booklet. . . . He couldn't stop talking about it, he was so thrilled. And that is just, like I say, one of many things I was doing. And this was off the chart. Who does this? That is the level of creativity that I have."

It was abundantly clear from talking to her that she was an extremely efficient and organized administrative assistant. But she wasn't better at being an administrative assistant than anyone else — she was just very competent. When she was told that, for the job market, she needed to come up with her superpower, what she was truly gifted at doing, she came up with the part of her job that no one hiring seemed to value enough to decide to hire her for that skill.

Some might say, then, that she should find another job — perhaps as an event planner or a party planner. But any event or party planner would point out the myriad ways in which planning an event is not simply planning an office birthday party at a different level of scale — it requires an entirely different set of logistical skills. And she did not want to be an event planner. She wanted to continue doing a job that she knew how to do well. But she was constantly being told that doing something well was not good enough — she had to be uniquely talented at a particular aspect. And organizing birthday parties was something she did superbly. Yet while this is the kind of work that is important to maintain goodwill in an office, it can not easily be turned into a metric that shows a clear improvement to how a business is run.

In short, being unique on the job market is not always a good thing, and why you are unique isn't always going to get you a job. You have to be unique in the right way — a standardized way of being talented at some set of tasks that most people must accomplish at a job and that companies value.

According to this logic, it is possible to be unsuccessful at convincing other people that you are the brand that you have chosen for yourself, and yet it is not possible to be wrong about what your authentic self is in the first place. To demonstrate this point, Pepper told a funny story about a coworker who wanted to be seen as knowledgeable and helpful but who hid under his standing desk at work. Pepper was concerned that Frank, a Google Analytics expert, didn't feel like a real part of her team, and she didn't know what motivated him. "So I was like: 'Frank, what do you love to do on the team? What is your personal brand?'"

Frank told Pepper that he loved to be the go-to person on the team for Google Analytics and answer whatever questions anyone had. But, according to Pepper, his personal brand didn't match what he believed about himself:

> Because this kid, let me tell you something about Frank. We have these standing desks at work at Google, and they are supposed to promote brain activity or burn more calories or something while you are working, I don't know. And underneath it he's put like a comfy fluffy chair. And he sits underneath his desk like this, with his laptop in front of him, and that is how he goes about his day. Which is like, you are the least approachable person I have ever met, and I'm not going to ask you a question about Analytics because you are sitting under your desk warding off questions! So we kind of talked about it, and I asked Frank, "When is the last time somebody asked you a question? Or asked for help about Google Analytics?" And he sat and he thought—"You know Pepper, you are probably right. I can't remember the last time I answered a question about Google Analytics."

After this conversation, he decided to jettison the comfy fluffy chair and set up a time for Analytics "office hours." This worked. Frank became the product guru of the team and one of its most influential members. For Pepper, the point of this story was not that her coworker was wrong about what his authentic self truly was, but rather that his actions at work did not encourage other people to perceive him in the way that he wanted to be perceived. When you work at aligning your branded self with your authentic self, you are trying to be conscious enough about all the ways in which you are representing yourself to others that you ensure that you are accurately conveying the message you wish to send. This, of course, implies that you have considerable control over how others perceive you.

Control

When workshop leaders talked about branding, they invariably started referring to how people presented themselves on the web (although you are also supposed to make sure that your offline interactions reveal your brand). As they talked about how people interacted online, they stressed that this was entirely and completely under people's control. Nick, the workshop leader at the San Francisco Public Library whom I mentioned earlier, asked people what concerns they might have about creating a profile. His audience told him that they were worried about being too public and about government surveillance. He responded:

> It's a two-sided coin. Yes, you are putting information out there for people to see and view. But if you think about it though, you also are in control of what you put online. Now certain sites, especially LinkedIn, is going to have a percentage to tell you to add more to your profile—it is 70 percent complete or something like that. They want you to put as much information up there as possible. But it is completely up to you what you want to put out there. That's the biggest thing to remember: the posts, the comments, the information where you live, your contact information, all this information is yours to put up on sites.

Nick was typical; other workshop leaders also would insist that as long as people think carefully about everything they put on the web, privacy concerns are misplaced. Instead of worrying, before sending every email, tweet, or status update, everyone should ask if the statement about to be made is consistent with their personal brand. Other scholars of the internet, such as Sonia Livingstone and Alice Marwick, point out that this logic places maximal responsibility on the shoulders of the user. As Sonia Livingstone puts it, users "operate within an environment that has been substantially planned for, paid for, designed and institutionally supported in particular ways, according to particular understandings of anticipated use and in order to further certain interests."[13] To claim that users are fully responsible for any information about them available online ignores all the many ways in which other users, other companies, or the very design of the internet's infrastructures all contribute to the information that circulates about people.

These beliefs about how personal branding works intriguingly shift how Americans tend to think about the kind of control authors or speakers have over what they have said. Linguistic anthropologists have long known that Americans tend to believe that the meaning of a statement is determined by what the speaker intended. For example, if Fred insults Betty by saying "You look great today, it is amazing what makeup can do," and he did not intend to insult her, should she be insulted? Betty might hear Fred as suggesting that she normally doesn't look very good and needs makeup to cover up her features. Fred might be intending to compliment her on her effort to look good, and affirming for her that the effort was successful. Americans will tend to say that Fred's intentions should determine whether Fred has in fact insulted Betty. If it was unintentional, Betty should overlook the potential insult and decide he didn't know better. But this is a cultural assumption. There are plenty of other cultures where people would tend to think Fred was insulting Betty regardless of what he thought he meant by that sentence.

Yet the logic of personal branding does not follow this American assumption that the author's intention is what should shape interpretation. When people find it difficult to make sure that their branded self aligns with their authentic self, this is largely due to the assumption that other people's interpretation of a person's brand is more accurate than what the person thinks it might be. You might not have maximal control over how people will interpret your statements or actions, and you have to constantly adjust your branding activities as a result. But what you do have maximal control over, under this logic, is saying something in the first place.

Branding Online

Up until now, I have been discussing the tricky work of deciding what your personal brand should be and presenting it as properly unique. In these workshops, you weren't just supposed to craft a personal brand; you were also supposed to be continuously marketing your brand through different websites. Once you have successfully reduced your complex personality to three or four context-independent qualities, how is that brand communicated online?

People face two central dilemmas when they try to manage the complicated balancing act of personal branding through various social media platforms. First, people struggle with being consistent across a range of media platforms, especially when these websites often encourage other people to contribute to a profile. No one on Facebook or LinkedIn is the sole author of their profile, although they can typically delete other people's additions. Besides that, being coherent has become a new challenge during job hunting because people all too frequently experience the internet as a database that compiles contexts, and not all internet contexts fit well with each other. Who people expect you to be on LinkedIn is often very different from who they expect you to be on Facebook. Second, people often struggle with the practicalities of acting like a business when they don't have all the resources and employees that a small business might have to market its brand, or the same forms of monetary compensation for marketing a self that businesses have for marketing a product.

It was clear in all the workshops I attended that once you figured out your personal brand, you were supposed to turn to social media to let others know what your personal brand was all about.

And when you did this, every single workshop instructor stressed, the most important thing to do was make sure that you were consistent. To create a personal brand, job seekers were told that they have to create a coherent self across a range of different media in order to give potential employers a sense that their internet search has turned up a reliably authentic self. And this sense of authenticity, according to workshop leaders, is produced by consistency: by making sure that you behave in a uniform fashion on Facebook, Twitter, and LinkedIn. Susan Chritton, a personal-branding strategist, explains how to fashion a personal brand: "From business cards to your website, you want to create a consistent visual image for your brand that makes the right impression on your target audience. You want to select images, colors and fonts that create the visual effect that expresses your personal brand."[14] Yet at the same time as you are creating a consistent brand impression across all forms of communication through, say, black-and-white images, you are also supposed to be a flexible self who can constantly enhance your collection of skills.

People are being told that one of the main tasks of having a personal brand is being consistent online, but being consistent is actually much more difficult than all the promoters of personal branding seem to assume. As people use Facebook, they often produce with the rest of their Facebook network a profile that serves as a complex map of who they are. This Facebook profile depends on who a person interacts with on Facebook—not only who a person adds as a Facebook friend, but who shows up most often in his or her Facebook newsfeed—and how those people like to use Facebook. And Facebook encourages people through its interface to represent themselves in different ways than other social media do. Your Facebook-specific profile can be different from your LinkedIn profile, your Twitter feed, your Pinterest profile, or your blog. You might interact with different people on your Facebook profile than on your LinkedIn profile. In fact, almost every job seeker I interviewed clearly kept their Facebook network separate from their LinkedIn network. Yet when people try to brand themselves on new media when applying for jobs, they are told to align all their different new media presences despite the fact that members of different social networks have different expectations of what people should say on a particular media, how often they should post, and, in general, how they should interact on that particular platform.

This dilemma occurs because every media has its own *participant structure*. This is a term that linguistic anthropologist Susan Philips coined, based on Erving Goffman's influential work in sociology, to describe the ways in which participation is organized in a given context.[15] Participant structure refers to the types of roles that are available in a situation and how people interact with those roles. For example, every workshop on branding has to have a teacher or two and students. There are some optional roles: sometimes the person who organized the event is present, and sometimes an observer is present, like an anthropologist or the teacher's boss who wants to make sure that he or she is good at presenting this material. In Ramina's context, there were also probation officers functioning as a security measure. Participant structures not only shape what roles are possible for people to occupy, but they also define how they are occupied. A participant structure shapes who speaks when—for example, in a branding workshop, the organizer (if present) will introduce the teacher, who then typically uses a PowerPoint and talks for most of the time. It shapes who listens and how—in the same scenario, what ways are listeners signaling attention or inattention, and are they involved in back-channel conversations, for example, scribbling notes to someone else in the room, or texting someone in a supermarket about whether there are enough eggs at home? A context's participant structure also determines how people are prevented from taking up roles or speaking—in the workshop example, how do you keep certain people from acting like the organizer of the workshop, or the leader? And finally, it refers to how the person who is speaking hands over the task of speaking to someone else. The teacher might ask questions of the audience and then call on people who have raised their hands or indicate their desire to speak some other way. This is a way of directing conversation that is specific to class settings. Think about how strange it would be to be in a casual conversation with a friend if every time your friend wanted to stop speaking and turn the floor over to you, he or she would point at you and nod, acknowledging you now had the floor.

Susan Philips developed this idea of participant structure because she wanted to figure out a problem that was happening in a face-to-face context. She was interested in why the Native American students[16] in schools on the Warm Springs reservation seemed to do reasonably well in elementary school classes but became disinterested in school by the time they were ten or eleven. After observing these classes for a year, she realized that these Native American students were losing interest in formal education when Anglo-American teachers imposed culturally inappropriate participant structures on them. She was able to explain how schools failed to educate these students over time as a consequence of how participants in a specific context adopted certain speaking roles, left them, invited others to take on certain roles, or prevented them from doing so over the course of years of clashing interactions.

One of her examples of a problematic element of a classroom's participant structure is that teachers on the Warm Springs reservation often required all students to seek permission to talk by raising their hands. For these Native American students, however, this was an inappropriate expectation. In the Native American communities she studied, only the speaker determines when someone should begin to speak. The fact that one person has stopped talking is not a signal that someone else should talk. In informal social contexts among Anglo-American speakers, the current speaker often determines when the next speaker can begin by ending his or her turn. When someone ends their turn, they also mark the end of their conversational turn by using certain intonations and word choices. Everyone participating in the conversation will share the tacit expectation that when the speaker stops, one of the listeners will then begin to speak. Not so in the Native American communities Philips studied, where people are content to let silence grow between one person's turn at speaking and another's.

This difference affected how comfortable Native American students felt in classrooms over time. In Anglo-American classrooms, during formal contexts, the teacher is supposed to control who speaks and when they speak. When Native American students were younger, the teachers organized fewer classroom activities in which the whole classroom was expected to listen to the teacher at once. Instead students played in small groups or worked on lessons on their own. As Native American students grew older, they increasingly stopped participating in classrooms because they were alienated by the Anglo-American teachers' expectation that a single authority figure should control communication in a classroom. Philips's account shows how very small changes in people's understandings of participant structures can have significant consequences. Native American students were stigmatized simply because they had different understandings of what counts as appropriate participation in a conversation. Different communities develop their own culturally specific understandings of how to speak and how to occupy the roles available in a situation, and sometimes even different understandings of what those roles are in the first place. And when these understandings clash, people can be excluded.

But what do participant structures have to do with personal branding? Personal branding is supposed to take place in a number of different contexts and using a wide variety of blogs and social media sites. Yet every technology configures the participant structure of a conversation differently, making it that much more complicated to create a consistent identity across these platforms. That is, every medium has its own structures that determine who and how many people can be the author of a statement as well as who is likely to be considered the author. In the case of a Facebook or LinkedIn profile, the profile is widely regarded as the offline person with an offline name and appearance that resembles the profile's name and posted images. Yet, as I mentioned, anyone in one's network might contribute to the presentation of self on the profile, thus a profile is a compilation of the statements of many people packaged as the profile of one person. With Facebook, the profile is quite literally a hodgepodge of many people's contributions. In addition, the medium also helps determine who can even participate in the first place: for example, Facebook does this by shaping who has access at any given moment, how and when posts enter someone's newsfeed, and how privacy settings are set up at a particular time. The medium also affects the perceived value of someone's participation and whether it is public or private speech. It affects whether participants in a conversation are hidden participants. It also determines how easy it is to forward a message to someone to whom it was not originally addressed. A medium encourages people to reflect on what is public speech and what is private speech, since how the medium is structured affects how many people can be easily addressed through that medium. For example, you might legitimately wonder when posting a status update on Facebook if it is, in fact, being read by only those people you intend to read the post. After all, people working at Facebook could be reading the post or it could appear in people's newsfeeds whom you don't know, because some of your Facebook friends wrote comments about the post.

The different participant structures create quandaries for users as they move from social media site to social media site. I am writing this in 2016, and at this moment, there are some social media sites that give people a certain amount of control over who has access to their posted words, and others that don't—a significant difference in participant structures. The logic people use to decide whom they will allow into their network differs from site to site. So undergraduates with whom I talk are occasionally confused when they join LinkedIn after having used Facebook for some time. They have already come up with clear guidelines for themselves about whom they will accept as a Facebook friend, but should these guidelines apply to whom they accept as a LinkedIn connection?

It is, of course, not just the expectations for interacting in ways appropriate to that specific social media network that cause headaches for people trying to fashion a consistent personal brand. Job seekers are also worried about making sure that their Facebook profiles will pass an employer's scrutiny—after all, anyone in that person's network can post something on the job seeker's wall, posts which may then be associated with the job seeker's profile. This is the downside of the logic of association that personal brands rely on. The fifteen undergraduates whom I interviewed were especially concerned about this, since the first two or three years of college were often a time when many people would accept any Facebook friend request if they had met the person briefly face-to-face. For some, the goal may even have been to have as large a group of Facebook friends as possible. It is only when they begin looking for a job that they start worrying about how their Facebook network shapes their Facebook profile, if only because they could be seen to be associating with others who have bad judgment about what they are willing to post on Facebook. People start to be concerned that their network connections could potentially threaten their presentations of themselves as employable on different social media sites. In general, job seekers can be flummoxed by the fact that their networks on different social media sites have site-specific expectations for participation, and this becomes a vivid dilemma when people try to present themselves as coherent, employable selves across a range of media, especially when what counts as appropriate uses are often still in flux.

This becomes a slightly different kind of dilemma when you are switching careers, and all your social media traces reflect the career you are trying to leave behind. Personal-branding aficionados tried to solve this dilemma by suggesting that people spend time curating their online associations and judiciously choosing which career skill set to mention. Vaneese, a career counselor and personal-branding expert, explained to me how this worked. We were talking about career transitions, and I mentioned that many people I spoke to found career transitions very difficult to manage, especially because their social media presence reflected the jobs they no longer wanted to hold. She lit up and said that she loved this dilemma. She knew exactly how to deal with it. She gave as an example a client she had recently been working with, a lawyer who wanted to start working in the entertainment industry. She recommended that he build his personal brand online by circulating articles related to his professional persona and adding a comment or two of his own to demonstrate he had unique insights: "Go onto Pulse, LinkedIn's curated news. Find articles and information related to entertainment. Then post and share those articles in your groups and on your page. And put a legal spin on it. Then people are going to start saying 'Who is this person, talking about law and negotiating contracts?' You are telling people the *what* and the *why* but not the *how*. The how, they got to call you."

By recommending that job seekers frame themselves as experts by recirculating articles, Vaneese is suggesting a form of expertise that is largely built on associations, mirroring other branding strategies that are supposed to work through similar types of associations. For example, Pepper, the Google recruiter, pointed out that in her LinkedIn profile, she made sure that the Google logo was prominently in the background. For her, this was a conscious strategy. "Just looking at this LinkedIn profile, what would you say is my personal brand?" Pepper asked. "Google," prompted one of the workshop participants. Pepper responded: "Google, right! I am leveraging another brand for my own personal brand. Why would I do that? It's not just to say 'Hey, I work at Google,' I promise. What do you think of when you think of the Google brand? It is innovative! It's fun! And that is the personal brand that I want to convey." Much of the way that personal branding is supposed to work is through these kinds of associations.

Developing a personal brand by recirculating various online articles works reasonably well for some jobs, and not very well for others. It works very well for people who want to be personal-branding experts—there are all sorts of articles about personal branding available that you can recirculate on LinkedIn, Twitter, or some other social media. If someone is working a retail job in a chocolate store, perhaps they can constantly comment on chocolate online. And even carpenters, plumbers, or electricians can do a version of this by circulating images of their successful projects, as I will discuss in chapter 6. But this does not necessarily make sense for various others, such as administrative assistants or security guards, for whom doing a job well is not about being an expert on a topic. In such cases, presumably, people would instead focus their branding efforts on promoting the three or four terms composing their branded self. Of course, there is the possibility that if everyone commits fully to using personal branding to get a job, then jobs that people have not typically associated with expertise will over time become associated with expertise. There may be a future in which bouncers circulate news articles with five tips for managing drunk customers.

The Labor of Branding

Yet even if someone is in a job in which the point of the job is being an expert, say, a research analyst, personal branding can be a burden. Tycho wanted to be hired as an independent research analyst, so he wasn't looking for a job—he was looking for a steady supply of clients. He explained to me that he had been working systematically at developing his personal brand for the past four or five years. He used LinkedIn and Twitter, maintained two blogs, and regularly wrote online articles about his area of expertise. He also routinely commented on published articles, getting into side conversations with the authors, if the topic was right. From my vantage point, he seemed remarkably successful at this, since his comments seemed to spark further conversations with people both online and offline.

Two days after I chatted with Tycho, in a completely different context, I met Tom, who, it turns out, also knew Tycho. He explained that he met Tycho initially through Twitter: Tycho kept retweeting his posts and then began asking thought-provoking questions about his tweets. Tom then met him in person after a talk that Tom gave at one of the Bay Area's many meetups. Tycho continued to ask his sharp questions at the talk, which led to subsequent conversations. I was impressed that so quickly after talking to Tycho about his personal brand, I stumbled upon this vivid example of how effective his strategies were.

Yet Tycho wasn't so certain: "I have kind of achieved a social brand, I guess, or a personal brand. . . . Well, actually, that's a very good question: have I or haven't I?" After years of maintaining an online presence, Tycho was starting to doubt whether it all amounted to anything, and he had a sophisticated take on whether his energy had produced measurable professional advancement. "I just read this paper from eight years ago by Baldwin and Carlyle," Tycho explained. "It's basically about *transfers* versus *transactions*. In an economic system, there are all kinds of transfers always occurring, but they don't always rise to the level of transactions." Transfers are any kind of exchange of resources or knowledge, and transactions are the special subset of transfers that include compensation.

In his estimation, all of his online "transfers" weren't becoming profitable "transactions." Sure, people were asking him to blog and share his insights on social media. In many respects he had become a successful "expert." But what did it mean? "There's a dead weight aspect to it that starts to develop, and it starts to take on a life of its own. . . . And in the early stages, it's kind of like, oh this is cool, this is great, I was on NPR marketplace (I was actually), but then it becomes [sigh]," Tycho explained.

Workshop leaders never acknowledged the practical dilemmas that Tycho is pointing to. Because a personal brand is so firmly based in other people's interpretations of you, it is always ambiguous whether you have succeeded in establishing the one that you want. Your brand might in theory be entirely under your control, but in practice it doesn't always feel like it. As Tycho points out, your brand "starts to take a life of its own"—you not only have to concentrate on managing your self, but you also have to manage your brand. Furthermore, it takes effort to maintain one's brand over time. Even if you may have established your brand temporarily, how long will this last? As Tycho explains, at first the signs of success are exhilarating. But over time, it sinks in that all the work you are doing to maintain your personal brand is free labor; no one pays you to do this. And then it becomes a practical question—has all of this actually been worth it monetarily? If you are supposed to think like a business, surely you should get the rewards of being one as well. In Tycho's case, he is grappling with the possibility that he could well have succeeded at being one of the world's experts on a particular topic, but this expertise is built on the uncompensated labor of personal branding. It isn't paying his bills or supporting his consulting business by landing him actual projects.

Tycho also alludes to another issue I encountered in other interviews: maintaining your personal brand takes effort on a weekly, even daily, basis. Other people have noticed this burden too. Alice Marwick, a scholar of new media, writes about how time-consuming personal branding was for Silicon Valley professionals. They often worked to promote their brands until midnight or 1:00 a.m. and felt generally overwhelmed by maintaining their social media presence.[17] You might want to treat yourself like a business, but you don't necessarily have the resources that a business has. One job seeker, a master's student at Indiana University, said it perfectly when he told me, "One of the reasons I wanted to mention the books being about small companies and start-ups, is that the suggestions in them are overwhelming for an individual. It's asking you to maintain social media constantly, answering blog posts regularly, and so on. It's too much for a single person, so I'm trying to find ways to make the workload manageable." He did not have the person-hours or labor that a corporation can draw on. One person is simply too limited by what a single body can physically do in a day to easily create an online presence that resembles a business.

Tarnishing Brands: Free Speech and Privacy

When people are businesses too, a good employment relationship is one in which both the employer and the employee have an equitable balance of risk and responsibility so that each party can maintain its own autonomy as a market actor. This is a pretty abstract ideal, but thinking about it in terms of branding can make this much more concrete.

Amanda is an airline flight attendant, and she wanted to be a pilot for Halloween. She is also a contract worker. She isn't a direct employee of the airline whose flights she services, but she still has obligations to uphold its social media policy. This makes dressing as a pilot for a Halloween party complicated. She could borrow a friend's pilot uniform, but then she would have to make sure that the airline insignia is masked, making her outfit a generic pilot uniform. Even with a generic uniform, she has to be careful never to be photographed with a drink in her hand or acting drunk, in case this appears online and compromises the airline's image. Fortunately she had had a lot of practice at avoiding being photographed at parties. She was on a sports team in college, and if she had been photographed drinking at a party, she would have been kicked off the team. Who knew that being on a college sports team was good training for the days when being photographed might result in being fired? What Amanda can do in her personal time is being affected by the company she is contracted to work with in the odd shell game that staffing agencies allow. The airline doesn't provide her with any benefits, but she has to mind its social media policies, even in her off-hours at a party. An employee's personal brand is understood to be an extension of the company's personal brand, and each has an obligation not to harm the other's image—however tenuous the employment contract might actually be.

Both the company as a whole and the employee are equally responsible for ensuring that their brand, their public image, does not harm the other. Employees believe that the company is supposed to maintain a reputation that will not tarnish the employee when the company only becomes a historical marker, a line on the employee's resume. By the same token, as long as the employee is working for the company, he or she has a responsibility not to tarnish the company's image in its public presentation.

This can have implications for the kinds of topics people feel free to comment on in various social media. It is an open question currently being adjudicated in the courts and by the National Labor Relations Board precisely what kinds of things people are allowed to say online about their working conditions. There seems to be a widespread legal understanding that employees should not comment on the company's products or customers' behavior online, but what employees can say legally about working conditions is still being determined at the moment of this writing.

More is at issue than just direct comments about the company, its practices, and its products. People might feel uncomfortable making strong political statements, in case the company they work for decides that this is a threat to the company's brand. While the internet was initially touted as fostering free-ranging political conversations, encouraging democratic debate to flourish, in practice employers' and employees' concerns about branding can interfere, inhibiting people from engaging in online political conversations.

I began this chapter with a question. Personal branding is an often espoused technique for job seekers, and yet no one I talked to on the hiring side seemed to care about applicants' personal brands—so why all the hype?

The primary reason personal branding has such appeal is that it is a logical extension of a new take on work: if the self is a business, then the self-as-business needs a set of marketing strategies, and personal branding supplies those strategies. Personal branding furthermore offers a solution to a dilemma contemporary employees face: how to present yourself as a desirable employee when you are frequently changing jobs and sometimes careers. In short, when your work history is no longer a coherent narrative, personal branding offers a (not so easy to implement) strategy for representing your self as stable and legible. But all of this involves ignoring context. To brand your self, you have to espouse a personality supposedly independent of context and do so across media platforms with different audience demands.

Given all the contradictions, why is personal branding so stubbornly popular? Frankly, it turns out that personal branding is a particularly useful set of techniques for motivational speakers and career counselors, but not job seekers themselves. In general, workshop leaders and motivational speakers face a problem—they are speaking to a mixed audience about how to get a job in a generic way, but all too often, getting a job depends on the specific idiosyncratic configuration of a workplace. Personal branding is a useful generic set of techniques if you are trying to give people good enough general suggestions because these are practices that anyone can do, regardless of what job they have. Recent parolees working at Walmart, classical musicians, or CEOs—all can have personal brands. From the point of view of people providing career advice, it is handy, indeed essential, to be able to offer to job seekers such a versatile and supposedly universally applicable form for representing an employable self.

Not all career counselors I met during my fieldwork talked about personal branding. The ones who did not were career counselors who worked for government-sponsored or community-based organizations geared toward helping unemployed people, although Ramina is an exception. When I asked these counselors why they didn't talk about personal branding, they said that their clients reacted strongly against the idea. They had mentioned it a few times as a potential technique when personal branding first became popular among the career counselors they talked to, but their clients thought it sounded too much like marketing, and they didn't want to be commodities. These counselors found other ways to help job seekers orient themselves to their changing job markets, often by focusing on networking.

Sometimes when I talk about personal branding to people who are unfamiliar with the concept, they will ask me: how is this different from having a horoscope sign or a Myers-Briggs type? After all, in the past, there have been many popular attempts to use a fairly limited set of terms to reveal what someone's personality might be. I hope that by now some of the differences are apparent. Self-branding is an instance in which you select the terms for yourself based on your self-understanding, and then try to imply those qualities through all your social interactions both online and offline. If you can't convince others indirectly that you have these qualities, you have failed to brand your self. It also involves specific techniques that were originally developed to associate objects to personalities, as well as social media practices that require a constant investment of time. It encapsulates tensions that are inherent to viewing the self as a business in the United States: You need to be both flexible and stable at the same time. You need to constantly improve yourself yet stay authentic. And staying authentic is not a given—you have to work hard to be your true self.[18] You also need to be both unique and predictable, and unique only in ways that lead to financial gain.

Finally, managing your personal brand requires that you constantly monitor how you are representing yourself as a desirable employee—there is no division between your work life and your personal life, or your work social media network and your personal social media network.[19] You need to treat every social interaction on any media as a moment to be true to (or potentially, to risk undermining) your personal brand. The sharp division between work and personal life under the self-as-property metaphor has evaporated. Now that you are a business, there is no break from being a business.

| Two | **Being Generic—and Not—in the Right Way** |

I have to admit that, after many months of talking to job seekers and hiring managers, I still have trouble answering the question: how do I get a job? From an anthropologist's perspective, there is no way to tell someone how to get a specific job. There are only suggestions I and others can make about how to *avoid* doing certain things that will prevent someone from getting a job. But actually getting that job? This is a mystery to many for some very good reasons, largely to do with the kinds of information that job applications provide to employers and how workplaces actually function when choosing who will join them. If the first chapter of this book is about job advice that I never found to be as effective as people hoped it would be, this chapter is about some of the problems with giving generic job advice in the first place.

This is not to criticize the many thoughtfully constructed workshops I attended about how to create a LinkedIn profile or how to write a resume. The workshops for job seekers were insightful glimpses into a complicated process, efforts to teach people the general outlines of how to produce the right documents and appropriate social interactions that could lead to a job. Being a competent job applicant goes a long way. But this advice can only take a person so far. Following the advice can position someone to get a job, but it doesn't ensure that they will get a particular job. And knowing about the advice is handy not because it is always such excellent advice, but because everyone else hiring and being hired has heard many of the same guidelines. If most people know the same rules for the game, when someone doesn't follow those rules, it is likely to be noticed.

Many of the recruiters and HR managers I talked to were contract workers themselves and had gone through the same short cycles of employment and unemployment as the job seekers I was interviewing. In fact, I wasn't always sure when I was interviewing a hiring manager, a recruiter, or an HR manager if this was an interview with a job seeker or an evaluator of applicants. Often I would begin interviews with hiring managers thinking I was going to ask a set of questions about how they selected good candidates, but the people I was interviewing mainly wanted to talk about their difficulties finding a job the last time they were looking. And as they described their job-searching experiences, it was clear that they too had heard the advice that was so frequently circulating in these job-seeking workshops and given by career counselors. If someone is violating a commonly discussed rule about how to apply for a job, the people evaluating that application are likely to spot the violation, and they could potentially use it as a reason to reject the candidate.

But this too is not guaranteed—some violations attract an evaluator's attention and encourage that reader to pay some more attention to the application. You simply have to violate the rule in the right fashion. What is the right fashion? It depends on the context, on that particular workplace. Here is one of the important ways in which, from an anthropological perspective, the job market is unpredictable. Everyone is engaging with standardized forms to represent themselves, but they have to use these standardized forms to represent themselves as distinctive. Figuring out what makes you distinctive is an act of interpretation, but how people will read your efforts to portray yourself as distinctive also depends on context-specific interpretations. After all, these forms are then evaluated by people who see their workplaces as distinctive and are trying to find people who can fit in with that workplace's complicated collection of people. I am going to discuss this dynamic at length, but before I can, I have to explain some concepts that linguistic anthropologists use to analyze social interactions like applying for jobs.

Genres for Getting a Job

If you go to any workshop for job seekers in the Bay Area, you will be told that applying for a job rarely begins with filling out a job application. There is much work that a job applicant has to do to be ready before beginning to apply. These days, if you want a job in the knowledge economy, you have to have business cards or an app on your smartphone that can function like a business card, you have to have a LinkedIn profile, you have to have a resume, and you probably should have a short, pithy way to describe what you do and what kind of work you would like to do. Interview answers too are something you are supposed to have already practiced,[1] and they are supposed to follow a particular format, what career counselors in the Bay Area called the PSR format[2]—an explanation of the problem, your solution, and the result. There is a repertoire of genres that allow you to present yourself as an employable person that you should be competent at producing before you even apply for a job.

It may seem a bit strange for some readers that I want to call business cards and resumes genres. After all, many people think of genres as a way to classify movies, plays, or books, even perhaps as a way of classifying stories that lets them know what they will like and won't like. Some people know that they don't want to go see *Annie* because they hate Broadway musicals. It doesn't matter what actors are in the performance or who directed it, if it is a film or a theater performance, or what the actual lines or songs are. They just know that they don't like watching anything that is a Broadway musical.

But how do they know that they don't like Broadway musicals? What is it about this category that allows people to predict whether they will enjoy *Annie*? They know that they don't like a particular genre because this is part of how genres function—genres classify how a certain story or an event will be structured. Someone might know that they don't like people bursting into song in the middle of spoken dialogue, a song that is often accompanied by dancing but whose lyrics rarely move the plot forward. Genres provide reliable routines even if they don't dictate the content. They are predictable forms for organizing how knowledge and experience are presented and circulated.[3] These ways of organizing an experience are predictable enough that you can know that you don't like an entire genre.

For linguistic anthropologists, genres do more than shape how stories are structured. The concept of genre can also be used to understand how people classify social interactions. For example, every month I go to a faculty meeting, a meeting which every professor in my department attends so that we can talk about how to run the department. This is a predictable event. I know that certain kinds of information will be circulated at the faculty meeting, and I know how it will begin and how it will end. I don't know exactly who will say what, but I know that there will be an agenda—a one-page schedule (sent by email two or three days before the meeting). This schedule will indicate to everyone the order in which one person will present information. After this person presents, then everyone in attendance may discuss the information, but they often, although not always, must indicate to the chair of the department that they wish to speak, and get acknowledgment from the chair that it is their turn, before they say anything. There are a number of other predictable routines that structure a faculty meeting. This form of organizing a social interaction helps make faculty meetings a genre. And this is a genre that I dislike. It doesn't matter what is being said. It doesn't matter who is in the room. I know before I even enter the room that I am going to heartily wish that I was somewhere, anywhere else for the next hour or two. I simply dislike faculty meetings as a genre.

Genres also strongly suggest ways in which social interaction should take place. That is, they presuppose participant structures, which I described in the previous chapter as the structures that organize the roles you can have in a given situation, that shape how you take on a role. Participant structure guides how you can encourage other people to take up specific roles, or how you can try to prevent people from doing so. It also shapes how you can change the roles you inhabit, or relinquish roles, or insist that other people stop inhabiting a certain role. The strong link between genre and participant structure is clear if you think about a phone interview for a job as a genre. Typically a recruiter or HR manager will set up a time for the interview by email or LinkedIn message. At the established time, the recruiter will call the job applicant to ask a series of questions. The job applicant has a sense of the kinds of questions that will be asked, because the applicant knows what the genre of a job interview implies. The questions will in some form or another address the interviewee's previous work history and how he or she has accomplished certain tasks in the past. There will almost always be a moment in which the interviewer will ask the applicant if there are any questions that he or she wants to ask. There are two roles involved—the interviewer and the interviewee, and the interviewer tends to control who speaks when, although the interviewer and interviewee collaborate on deciding when certain information will be revealed.

Yet this participant structure can be tweaked. I heard an illustrative example of how a job seeker can shift the participant structure in a phone interview when I was attending a networking event for job seekers. I was trying to explain to the group the argument I spell out in chapter 6 about how hiring managers, recruiters, and HR managers don't always have the same goals when interacting. A conversation ensued about how the present job seekers deal with the ensuing tensions, tensions which for them are often behind the scenes. Job candidates may know that these tensions exist, but typically they won't get to see any visible signs of trouble during the hiring process.

Phil started to describe his technique as an interviewee for dealing with recruiters in a phone screening interview:

Phil: One of the strategies that I use is to help educate the recruiter, and then the hiring manager themselves, about: "Tell me a little bit about what you are really looking for and what you are trying to do." And if they didn't know that, then I would give the recruiter questions to go back and ask the hiring manager, which in turn they would say "Well, wait a minute, yeah, okay, I want to talk to this guy."

June (workshop leader): Being on the same side of the table versus the other side, very smart.

Phil: Yeah, you have to put yourself on the other side of the table and you have to ask questions—"what are you really looking for?" Because I can tell you from hiring myself, and I am sure that all of you around the room have had this experience, there are a lot of cases: Here's a job description. And when you talk to the hiring manager you wonder who wrote this job description. That isn't what they are looking for.

Although she understandably doesn't talk about it in terms of participant structure, by talking about switching sides of the table, June points out that Phil has consciously changed the interview's participant structure. Phil is asking questions that shift the recruiter or HR manager from being an evaluator to being a mediator or, more accurately, a slightly inadequate representative of the hiring manager—a slightly inadequate substitute for the hiring manager because he or she can't in the moment anticipate what the hiring manager might say. The recruiter then will promise to act as a mediator, announcing that he or she will convey Phil's questions to the hiring manager and report back. Phil even points out that maneuvering the recruiter into asking the hiring manager these questions often means that he will speak directly to the hiring manager more quickly than would usually happen in a job search. He is also bringing the job description, and in particular the unknown and not very careful author of the job description, into this interaction. At the same time, he is slightly shifting the role of the recruiter from representative to translator, getting the recruiter to act as a translator of the hiring manager's desires. He points out that the recruiter may not be the author of the job description, and indeed may be having trouble interpreting what kind of job applicant the author of the description does want to hire. By invoking the author of the job description and maneuvering the recruiter into contacting the hiring manager by "being on the same side of the table," Phil is doing quite a bit to change the phone interview's participant structure.[4] As this example shows, even though genres provide a framework for how interaction will take place, any participant can change the structure of the interaction to a certain degree in the moment, and thus change what happens next. However, to change an interaction in a direction you want, it helps to understand the basic structure of the genre and know where there are opportunities to change patterns of interaction effectively.

In general, to be hired, job applicants have to be reasonably competent at a range of genres. They have to fill out online job applications correctly, have a presentable resume, create a LinkedIn profile that recruiters will find by searching keywords (depending on the job), and so on. While readers might expect resumes and interviews to be necessary genres that a job candidate must be competent at, there are other less likely ones that can also be equally relevant. For white-collar workers, business cards are often a crucial genre in this process, because you need a way to exchange contact information easily at networking events or after unexpectedly promising conversations in a supermarket line. It isn't necessarily the actual paper cards that matter. Some companies have tried relatively successfully to develop technological substitutes for business cards. For example, smartphone owners are able to avoid business cards entirely by using apps which allow people to exchange contact information by "bumping" their smartphones together. This, of course, requires that both people own smartphones with the right apps.

Having a business card can indicate a person's familiarity with the job search process. People new to the job-seeking process might not have a business card for the first couple of weeks. After all, if someone has been recently laid off, they may have many business cards, but none of the cards will have the right information—they will all have the contact details for the previous job. The job seeker has to remember to order new business cards and figure out how to describe him- or herself on these cards. This was actually one of the ways while doing research that I could figure out relatively quickly whether someone I was talking to was new to a job search—if they didn't have a business card, they had probably only started looking actively in the previous week or so. One man told me how friendly people were at a meetup he attended when he first started looking for a job. Why did he think they were so friendly? When a man started asking him for his business card, he awkwardly admitted that he didn't have one. He had only recently been laid off and hadn't gotten around to getting his own business cards that didn't mention his former company. The man he was chatting with told him not to worry and found some sticky notes for him to use. He spent the rest of the meeting handing out his contact information on those sticky notes, and no one seemed to mind at all.

In this example, sticky notes could substitute for business cards only because what is important about business cards as a genre is how they allow knowledge to circulate. The shape and size of the business card genre affects how the interactions at that meetup will be taken up again later, after the event. You can put someone's card in your wallet or purse, tucking it away for later use, and any piece of paper can serve this function, even a sticky note.

Genres not only shape how information is organized in the moment of interaction, but they also shape how the knowledge discussed in one interaction will be able to travel into other interactions—maybe on a small piece of paper, but also maybe by being told in ways that make it relatively easy for someone to share the information with someone else afterward. For example, some people would tell me that what made a good answer in the first screening phone interviews was a concise answer. They thought that being concise was especially important because of the nature of this type of interview. Apparently recruiters or HR managers could take notes more easily when listening to a concise answer and then bring that information to the hiring manager for consideration. This is an example of how genres anticipate the ways that parts of conversations may be separated from the context in which they are produced and repeated elsewhen and elsewhere.

Genre Repertoires

Job seekers, in order to be hired, are never using just one genre; they are producing a genre repertoire.[5] Even many job descriptions are clear that applicants should consider submitting more than one genre that provides evidence of employability—after all, resumes are a different genre than cover letters or job application forms. Indeed, almost every workshop for job seekers I attended was focused on helping people figure out how to be competent in at least one of the genres that those hiring used to evaluate job applicants— workshops on, for example, how to use LinkedIn, how to conduct an informational interview, how to write resumes and cover letters, how to interview, and, finally, how to negotiate an offer. There are other genres that people might use in the process of trying to get a job—the job descriptions that people read, the Excel spreadsheets that some people make to track what happens with each of the applications, the notes the recruiter may take during a phone screening interview, the evaluation forms a hiring committee may fill out to assess each interviewed candidate. For now, I will only discuss the collection of genres that job applicants use to present themselves as employable, the genres that both those trying to be hired and those seeking to hire will see.

In the hiring process, all the many aspects of the different genres that people use intersect with each other in complicated ways. The differences between resumes and LinkedIn profiles illustrate this well. In my introduction, I suggested that at a first glance, there may not seem to be much of a difference between a LinkedIn profile and a resume. Both are, as many people told me, "marketing tools" for an applicant's work skills and professional history. In calling resumes "marketing tools," people are tacitly referring to one of the ways in which the metaphor that the self is a business is now commonplace.

Yet resumes and LinkedIn profiles are not the same genre of marketing tools, or at least, not in the way that linguistic anthropologists are defining genre. In applying for a job, resumes are sent by email or, much more rarely nowadays, on paper, to a specific workplace. This gives applicants the chance to tailor a resume for a specific job in that particular company. The resume does not address the widest public possible. Instead, these days, the submitted resume is supposed to be rewritten, from a person's generic template of a resume, for that particular job, and thus for the handful of people at a specific company who will be selecting job candidates. By contrast, LinkedIn profiles are public, online, and composed with the broadest professional audience in mind.

Many people would tell me that describing resumes as a tailored marketing tool was a change in how people thought about resumes over the past ten years. They would also say that the ways in which people put together a resume had changed as a result. People would explain that previously, before the 2008 crash or the 2000 dot-com bust, resumes were supposed to be a straightforward record of one's working history, although according to other research on unemployment, this change happened earlier, perhaps in the late 1990s. You were supposed to list chronologically where you had worked and describe briefly the skills you used at each job. And importantly, you were supposed to list every job you had had—as a job seeker, you only left a job off your resume if it might genuinely harm you to allow potential employers to know that you had had this job, for whatever reason.

These days, applicants are strongly encouraged only to put down the jobs that are relevant for that potential employer—not every job in someone's past should be listed.[6] This change may just be practical. Nowadays, people may have too many jobs to comfortably fit them all on a one- to two-page resume, especially as more and more people freelance or have contract jobs for short periods of time. Some job seekers also may not want to list all their jobs because this will signal to an employer that they are over forty, and they are worried about facing age discrimination.

Even with all these possible reasons, I most often heard one explanation for why someone shouldn't list all their jobs on a resume— not all jobs are relevant for a specific job description. One instructor in a resume workshop explained how to tailor a resume, stressing the importance of relevance with a funny example.

"I once climbed Mt. Kilimanjaro while I was knitting a sweater for the Dalai Lama," she said, feigning pride. "It was quite an amazing accomplishment! Right? And I made a lot of connections in that country, and picked up some of the language. Wow, you can't even believe how amazing I am."

"Oh, and by the way," she said, with precise comedic timing, "I am applying for a job as a grocery clerk. . . . They do not give a damn that I climbed a mountain and knitted a sweater! It is not related to the job. Don't put it on there, it is a distraction."

While resumes can be tailored, it is hard to use the same strategies for a LinkedIn profile. If you are a job seeker, your LinkedIn profile is supposed to be public so any potential employer or recruiter can find the profile easily. And herein lies the problem. LinkedIn profiles, because of how public they are, must be written to be as general as possible—to allow you to connect to as many jobs as possible. This can involve including details that resumes, limited as they are to one or two pages, don't contain. If you have in fact managed to become the flexible, multiskilled self that the self-as-business logic wants you to be, you risk seeming too incoherent, too scattered on your LinkedIn profile. If you genuinely are flexible as a job seeker, and willing to take a range of different jobs, providing supporting evidence for this on your profile could mean that recruiters or hiring managers might think you are unfocused or difficult to classify, should they choose to interpret your profile that way.[7]

The ways that a LinkedIn profile differs from a tailored resume can cause problems in another way. A tailored resume is supposed to circulate among a small group of people within a certain company. All the work a job seeker has devoted to imagining a potential relationship with a specific business can be unraveled when the expert reader—the recruiter, HR person, or hiring manager—compares LinkedIn with the tailored resume. And he or she often will. One woman explained in a workshop on LinkedIn profiles how this became a concrete dilemma for her that she was still puzzling over:

Maria: I have a question about this—one time one HR manager looked up my LinkedIn profile, and she had my resume too. And she said it should match *exactly* my resume, and it doesn't because I worded it different. I didn't want it to match actually. So I don't know what people think about that, if they have any feedback on that. I didn't change it, because I still don't think it should match, but I don't know. That's what she told me. . . . But I don't see the point. If you have my resume, then you don't need my LinkedIn profile, because it is exactly the same. I don't know.

John: It certainly shouldn't conflict. . . . But she didn't say it conflicted, she just said she wanted to see both the same.

Andrew: I mean, in your situation, she wanted to see an exact match, but I bet a lot of recruiters are interested in finding out more about you, what's up with you.

The job seekers in this workshop were puzzling over the practical problems that emerge when using two different genres that are supposed to do similar work in a context where people can have contradictory interpretations of what the gap between these genres is supposed to be. They were discussing the fact that the people evaluating applications often have different views of the connections between both genres, and were debating how seriously to take one HR person's strategies for sifting through applications.

Here is a moment in which being flexible, which is supposed to be an attribute of an ideal worker, can come into conflict with the confusing work of representing yourself as employable through both a resume and a LinkedIn profile at the same time. In fact, this is a moment in which these two genres are seen to conflict precisely because of today's job market's contradictory messages on being flexible *and* well-qualified for a specific job. Often LinkedIn profiles are viewed as encouraging you to be as general as possible, so that you show precisely how flexible you can be. This is especially true if you are applying to more than one type of job at the same time—you want the hiring evaluators from either kind of job to be able to see your LinkedIn profile and think you are a good fit for their job.

While this might sound like an abstract quandary, in practice, people will talk about how to deal with this by focusing on formatting and where people should place certain information on their

LinkedIn profile. I would occasionally ask people to read LinkedIn profiles with me and explain to me how they were interpreting the information they saw. Nico, the CEO of a marketing firm specializing in sustainability initiatives, showed me the profile of a friend of his who was in the middle of a career transition (perhaps, but she thought she might return to her former career if she got a decent job offer). "And this part I just don't get at all," he said. "She basically says one six- or seven-word sentence about real estate and then immediately lists that she is a marketing professional and has expertise in digital media."

Nico was not sure his friend had successfully talked about her previous career in a way that highlighted her transferable skills: "This to me is one of those tough things where you are making a career change and how do you take advantage of leveraging? . . . She's not doing that but she is still trying to claim something. . . . And then the whole formatting is just off."

Nico is commenting on the problems that a LinkedIn profile can pose to a job seeker when that person has had too many different types of jobs. Of course having too many different jobs could also be seen as being the self-enhancing, constantly learning worker who is supposed to be the current ideal. With resumes, you can be specific enough that skills you developed at other jobs that are irrelevant for one type of job don't creep into your representation of your past job experiences. But the generality of the LinkedIn profile is viewed as undercutting the specificity of the resume.

LinkedIn profiles are too general in another sense as well. You often want to have as many connections as possible, and often you have been building up these connections through several job transitions, and occasionally even wholesale career transitions. This can make the different professional networks that you have seem visibly incoherent to another person who is evaluating your LinkedIn profile by interpreting your connections. Although I do want to point out that this was a worry I heard job seekers express but never heard of anyone on the hiring side actually doing.

In general, differences between the genres that people use to represent themselves as employable can create dilemmas for job seek-ers, just as the differences between social media platforms can be an obstacle for people trying to craft their personal brand. The ten-sion between resumes and LinkedIn profiles occurs for a couple of reasons. First, these two genres presuppose different participant structures. A resume is typically written for a much smaller group of people than a LinkedIn profile, although resumes posted on job sites are different. Second, when the self is a business, while the ideal may be that people constantly enhance their skills and experiences in as many ways as possible, this very flexibility can be a problem when you are trying to persuade others in a company to hire you. Not all genres encourage the right mixture of flexibility and specificity for the complex social task of being employed, and just as importantly, not all forms do so effectively when they are combined together. The flexibility that may seem to be an ideal when you read advice on the web or in books about how to find a job can be an obstacle in prac-tice. And sometimes flexibility only becomes apparent as a prob-lem when you start dealing with all the different genres you have to use to present yourself as employable as all part of the *same* genre repertoire. As long as you don't have to worry about the same per-son looking at your LinkedIn profile and your tailored resume, you might not worry about having to choose between being flexible or being readable when creating these genres.

The Hourglass of the Job Search

Among genre repertoires in general, the genres collected for job searching tend to be geared toward standardization, toward encour-aging everyone to represent themselves as unique and distinctive by using routine forms. These forms, in fact, are so codified that there are easily found templates available for each of these job-searching genres. You can find hundreds of slightly different templates for how to compose a resume, for example. This is not a new problem, which became apparent when I was reading job-advice books on cover let-ters and interviews from the twentieth century onward.[8]

As a result, I have come to think of the job search in terms of an hourglass. The top of the hourglass represents job seekers. They are people with remarkably complex social relationships and life experiences that they try to reduce and simplify into forms, like resumes, that provide a very restricted view into a person's complexity. The genre repertoires for job search serve as the constricting tube in an hourglass. These genres reduce all the vagaries and depths of a person's life experiences into the most telegraphic of glimpses into what someone has done with their days in the past and could do with their days in the future. A job seeker sends this simplified summary of his or her work experiences to different workplaces (the bottom of the hourglass). But each workplace is complex and rich in its own right, filled with its own social dynamics. As anyone involved in hiring knows, selecting a person to hire involves coordinating a group of people, each with his or her own strategies for interpreting the documents and selecting likely job candidates. This evaluation is done in the middle of all the intricate coordinated and uncoordinated social interactions of a workplace.

The genres that people use for connecting these complex people and complex workplaces together are genres geared toward reducing complexity and erasing nuance. These are genres that have come to be adopted and circulated widely largely because they represent complex human beings in such reductive forms that job applicants can be compared with one another. And while sometimes hiring managers and recruiters will comment that comparing two applicants is like comparing apples and oranges, these forms have already done enough of a disservice to these applicants that they end up all looking like squashed fruit.

I always found the people I was talking to much more interesting and unpredictable than any of their LinkedIn profiles or resumes might suggest. But, let's face it, no one I talked to worried much about the price of reducing their own complex life history to a professional representation. Everyone seemed to take this task for granted. They mainly struggled with how to do it, not questioning whether it should be done in the first place, which is fair enough. Yet there were some complexities about their experiences that they were very conscious did not fit easily or smoothly into some of these genres, and yet had to be addressed in one way or another. People were concerned about how to finesse career transitions, how to move between countries, and how to return to work after being unemployed for a long period of time (and there were many different reasons why they might not have worked for a while). Transitions between careers and countries each presented their own challenges to the ways a standardized genre repertoire is supposed to allow evaluators to compare job candidates.

Recruiters would tell me that managing a career transition has become more difficult for people because hiring managers have become more demanding about the kinds of skills they would like applicants to have. They would say that the set of skills hiring managers now require has become more extensive. Recruiters were frustrated that if applicants haven't demonstrated that they have done a particular job beforehand, and haven't demonstrated all the desired skills from A to F, then hiring managers will reject the candidates out of hand. This doesn't always happen, but it is a common enough complaint that recruiters have about hiring managers—they have grown much pickier in the past ten years or so. One recruiter, who had been working for over twenty-five years in the Bay Area, explained: "In the old days, you know, if I was looking for somebody that had A, B, and C and they're smart, and they could pick up E and F, they'd get hired. Over the last ten years, five years it's gotten much, much more . . . they better have A, B, C, D, E, and by the way, F and G would be nice too. So it's a lot harder specs to get people hired. . . . A lot of them have an idea of 'I only want people with good pedigree. I only want people who have done a certain type of technology. I only want people from A, B, C, D, E, F, G, H, I, J companies. I only want people who've worked at a start-up before. I only want people who—and it's highly illegal but—who are youthful.'" Recruiters were quite reasonably frustrated that the hiring managers they were trying to satisfy were developing narrower and narrower sets of requirements. Indeed, hiring managers might ask for a person with five to seven years of experience doing a particular task, something less likely to be possible if someone is truly the ideal flexible self and switching jobs every two or three years. Recruiters and job seekers talked about how it has become much more complicated to present yourself as having transferrable skills. Skills that you use to accomplish a task in one type of job may not easily be interpreted by those hiring as relevant for another context, even if they seem like the same skills to you.

This was often true for working-class jobs as well as white-collar positions. You might think that being able to work one retail job means you could probably work any retail job, regardless of what you were selling. But this turned out not to be the assumption that hiring managers had. Monica, who hired people to work behind the counter at a small chocolate shop, felt that selling food required both some previous experience and an ability to describe food poetically to customers. She explained to me what questions she likes to ask in the in-person job interviews, explaining that she always asked applicants to talk about the best meal they had ever had: "Cause I wanna hear how they describe food. Because we have to describe the chocolates. We have to describe pastries. It doesn't have to be perfect, but I don't want them to say, 'My grandma made a lump of mashed potatoes and there's butter.'" Monica is discussing one of the complex skills necessary to accomplish something as mundane as selling chocolate—you have to be able to translate your taste sensations into a verbal description that persuasively communicates to someone else whether they would want that sensory experience.

People often mention the problems of shifting careers in the same breath as they describe an increasingly shrinking tolerance for training people on the job. Both job seekers and recruiters told me time and time again that companies no longer were willing to train people for a job, that they wanted people to hit the ground running.[9] This is a consequence of changing the hiring relationship to metaphorically being a business-to-business contract. When a business decides to enter into an alliance with another business, it doesn't want to wait for the other one to develop into the right kind of partner. A business wants to enter into an alliance with another business that is well-equipped from the outset.

78   This, however, is a moment where the metaphor of self-as-business does not adequately capture what it means to enter a workplace. As Tom in HR explained: "The first year is not a waste, but you don't know what you're doing for the first 6 months and then you figure it out, that's the next 6 months. That's the first year. Alright, so that's a pretend first year. Now you have a business cycle. The second year you can really do something to see how it came out. I would say the third year you've made improvements on that, and how did that turn out?" Tom is describing how difficult it can be to become familiar with the nuances of a particular job. Workplaces are not uniform enough that one can easily and smoothly slot into an office in a month or two. You have to learn all sorts of details about both how to get anything accomplished within an office and how to interact effectively with people in other offices and other companies on behalf of the company you are working for.

Career transitions were not the only moments in which people had trouble representing themselves effectively in resumes or LinkedIn profiles. People also had difficulty moving between countries. Some of the problems revolved around credentials—regardless of your medical training overseas, in order to be a doctor in the United States, you have to take the US Medical Licensing Examination and, depending on the state and your specialty, be a resident for at least three years. Yet the problem is not just one of credentials but also one of classification. Audrey explained to me that she had just moved back to California after working in South Korea for a number of years for a Korean company. Since moving back to the United States, she has had trouble explaining her job effectively on her resume. The range of her job duties doesn't match the range ascribed to any single job in American companies. How people classify corporate jobs in Korea is different enough from how corporate jobs are classified in the United States that Audrey was stymied, and she didn't know what jobs she could convincingly apply for in the Bay Area.

   This problem of classification also exists for people trying to move between companies, especially companies that were established in different decades. Susan had been working at Hewlett-Packard for twenty-one years before she took an early retirement deal. She wasn't ready to retire, so she was applying for jobs as a project manager, the job she had at Hewlett-Packard. Yet she was finding it confusing to interpret the job descriptions and to explain to recruiters what she could do. What it meant to be a project manager at Hewlett-Packard was not the same as what it meant to be a project manager at other companies. She had coordinated projects, but she had never interacted with the factories overseas to ensure that they were creating products of the appropriate quality. The supply team at Hewlett-Packard had done that instead. What a project manager was supposed to do has changed over the decades, and a company founded in the 1970s asked project managers to do a different set of tasks than companies founded in the 1990s. So it was particularly complicated for her to persuade people to hire her, because the tasks she had done didn't always fit the titles she was applying for.

   Sometimes the problem did not revolve around how a particular company divided up the occupational roles. Peter explained the trouble he had making the transition from hardware marketing to software business development because the term *business development* refers to a different enough set of job functions that similarly named jobs for selling both hardware and software don't involve the same work at all. It took him awhile to realize that because of his experience in hardware business development, he was used to finding new accounts and developing a relationship with clients directly, yet in software business development, one develops strategic partnerships with companies who then sell through third parties.

   For both Susan and Peter, the problem was one of classification—companies or different professions classified the tasks belonging to a job position differently enough that one couldn't make effective comparisons.[10] Yet, as I mentioned earlier, part of the point of the resume or LinkedIn profile is to be able to reduce complexity enough that this abstracted comparison would be possible. Peter felt that this was not a problem he could solve only by learning the new terms for this profession. After all, he listed on his resume his previous job titles which included the phrase *business development*, and he had to get past applicant-tracking systems, and possibly also recruiters. "I got tripped up with this keyword thing," he explained. "The jobs I was going for had 'business development' in a lot of my titles. . . . So I was using that title over and over, and I think [my application] was caught because of this. The applicant tracker system at a software company sees 'business development' and kicks you into a different pool."

How to Give Advice in a Nonstandard World

The career counselors teaching the workshops I attended were aware of the problems of these standardizing forms. Career counselors are helping people wrestle with the complexity of their work experiences and reduce them into a form that can be easily and quickly read both by computerized applicant-tracking systems and by overwhelmed recruiters, HR people, and hiring managers. Ideally, these forms will be compared favorably to those of other applicants for the job. The counselors focused on explaining to the workshop attendees how to produce the genres correctly enough to pass these various screenings. At the same time, they were teaching people from a wide range of backgrounds, and, as they would occasionally remind the attendees, each profession has its own vagaries. So they were trying to teach these forms in such a way that people would know how to adjust them to fit the particular demands of the kinds of jobs they specifically were looking for.

The workshop leaders were also trying to prepare people to appeal to workplaces which had their own distinct techniques for selecting people. Workplaces are as complex and distinct as candidates. What would make someone a successful or appealing job candidate for one company might lead another company to reject the same person. Once the candidate can demonstrate a relative competence with the required genres and prove that he or she has the necessary background to do a job, then the candidate has as good a chance as anyone else similarly qualified. What makes one candidate more appropriate or attractive than another is up to the peculiarities of that workplace.

I heard a number of stories that convinced me that workplaces are distinctive enough that there is no good answer anyone can give to the question of what works when looking for a job. I talked to one HR manager who told me that people cared tremendously about eye contact at her company. "Our team is really harsh on that," she said. "I am a good critiquer too, but I give people a little grace for being human, maybe because it is the nature of my role. But I mean, [my team tells me someone is a bad fit because] 'he looked around the table, and then he looked at the ceiling.' Oh my God—please tell job seekers, they don't have to stare, but they need to make eye contact. . . . It is amazing, the nonverbal stuff that comes out in my debriefs." No one else told me that eye contact was important, but coworkers in that HR person's company had decided that someone's eye contact indicated something much more substantive about how the applicant was going to behave socially in the workplace. In this instance, people in this workplace, and this workplace alone, had chosen to read a specific physical gesture and decided to use this to predict a much more complex way of interacting that might or might not be accurate.[11]

People might have contradictory takes on similarly minor practices—which leads me to believe that once you have become competent in the genres, there is still quite a bit of uncertainty about how others will interpret what you produce or how you perform. For example, I heard contradictory perspectives on whether people should be concise in their answers to job interview questions. At one workshop, the leader stressed over and over again how important it was to be succinct, giving a vivid example of an interview disaster: "I did have a hiring manager say to me, '[This one time a candidate] went on to give me his *entire* resume. He talked for almost 20 minutes. At that point, we were done. I didn't want to hear from him anymore.' That's not a pitch, right? That's called burning out the hiring manager." Two weeks later, I was talking to a hiring manager about his pet peeves for how people respond to job interview questions, and he explained how frustrated he gets when people are *too* concise: "He was too short, he didn't give me enough details. And sometimes I'll go back and I say I want more details, tell me more about it, and they focus on being concise. And sometimes that's just not enough. You need to listen I think to what people are asking or saying. If I'm asking you to give me details and you're concise, you're not listening." This hiring manager wants far more details when people answer his questions, and others want examples of short, specific summaries. Neither is right or wrong, and intriguingly, both agree that conciseness is enough of an issue that they have careful analyses of how it helps or harms an interview—their analyses just happen to be contradictory.

At the heart of their disagreement are differences in language ideology—differences in how people interpret the attributes of a conversational turn or a way of speaking and ascribe a different range of associations to that practice. Language ideologies are the beliefs, attitudes, and strategies that people have about how language works. Linguistic anthropologists know that this is not the same as how language in fact works. Language ideologies shape people's linguistic practices but do not determine these practices. For example, people often associate elaborate personal characteristics to the way someone speaks. The elongated vowels in an American southern accent can be interpreted as "lazy" speech which in turn contributes to stereotypes of southerners as laid-back and easygoing. Or, another example, people who wait seven seconds before uttering a sentence might be seen as thoughtful, unwilling to say anything that they have not carefully considered. Waiting seven seconds before saying anything—who knows what inspires someone to do this? Habit? Family background? But not everyone shares this language ideology that a pause before speaking is a sign of careful intelligence. Some people view those who wait seven seconds as dull or not very bright, as slow-witted both figuratively and literally. The same is true of people's responses to concise answers in a job interview: conciseness can be interpreted as indicating something fundamental about someone's personality, but what that fundamental thing might be can differ from person to person or workplace to workplace.

These differences in interpreting signs or practices can become crucial, and also can reflect a transition in how people think about the hiring relationship these days. For example, there are a range of opinions about how long one should ideally stay at a job. In Silicon Valley, recruiters would often tell me that a person with the ideal career trajectory was someone who worked two to three years at a number of jobs, carefully choosing which companies they worked for. One recruiter, June, told me that she counsels people not to stay in one place too long. "Well, I mean it's a little bit too much to see two, two, two, . . . two years everywhere," she explained. "I don't want to see that. Because how much of a contribution did they make? But definitely don't stay there five, six, seven or more years. Because they become institutionalized."

This is a not-so-surprising consequence of thinking of yourself as a business. In the mid-1980s in Silicon Valley, companies quite consciously decided that company loyalty could no longer be the coin of the realm, or anything that they wanted to foster. While corporations began to actively encourage workers to see themselves as working on specific projects that might have a natural end, this concept took awhile to filter down to how people were evaluating job applicants. Hiring managers continue to be concerned about hiring job hoppers, people who make too many rapid transitions between jobs.

But what counts as too rapid a transition between jobs is still up for grabs and, moreover, seems to vary from region to region across the United States. For example, Peter found that what he considered a typical career trajectory was seen as a red flag for a company in Chicago. He has been working in northern California for years, and he found a job in Chicago for which he was willing to move. In the interview, he kept getting asked why he hadn't spent longer at different companies—why only three to four years? He tried to explain that staying five to eight years at a job, as people are expected to do in Chicago, would have been held against him by Bay Area recruiters.

And indeed, Bay Area recruiters have told me that people in other regions stay too long in jobs, making it difficult for recruiters to convince hiring managers that these applicants have initiative. Peter and the recruiters I spoke to felt that this was an example of regional differences, although I had heard stories of hiring managers in the Bay Area who also were concerned about job hopping when reading the resume of a job candidate who moved every two or three years. Here is another example of contradictory interpretations of the same information—people know that the length of time a person has spent at previous jobs is a potentially significant factor in evaluating that person as a job applicant, but what precisely that length of time means varies from evaluator to evaluator.

Workplace dynamics can often be too specific to a particular workplace for standardized advice to work all that well. One person in HR explained to me that she had come across the phrase "May the Force be with you" in a resume a few months earlier. I expected her to use this as an example of how a job seeker had torpedoed an application. But she said she laughed when she saw the resume, and explained that going through resumes was a boring enough task, and this at least made her sit up and pay more attention to the resume. She said if the job had involved marketing, she might not have been willing to forgive such an inappropriate presentation, but for the particular job she was screening resumes for, all this quirky gesture did was inspire more careful attention. She thought people at her workplace would appreciate this glimmer of humor, so she showed it to the hiring manager, who also thought the gesture was a great sign that the candidate would fit in well. Yet when I tell others this anecdote, they tell me that they would have dismissed this resume out of hand without thinking a great deal about it. From their perspective, this job candidate did not seem to understand that resumes are not an appropriate genre for being too quirky, and this inability to discern when it might be appropriate to be unusual is not a good sign, and might indicate that the candidate will be too socially unpredictable to get along in the workplace. This applicant was lucky that the screener did not think this way and instead was charmed.

Matt, a job seeker, told me a similar story about a Skype interview he had. He was sitting in his living room, and one of his cats kept walking back and forth in front of the monitor. He kept trying to shoo the cat away discreetly, since he did not want to call the interviewer's attention to the fact that a cat tail was often in the frame. Finally his interviewer asked if what he was seeing was in fact a cat. "I said, 'Yeah, that's one of my wife's cats.' And he says, 'Oh I have four cats!' Well, so do I! And so that went from what could've been a negative kind of thing to a bonding kind of thing. I picked up my cat to show him, and the guy says, 'Oh I had to lock my cats out of the room otherwise they would've been in the picture.' I went, 'Well I didn't think about that.'"[12] A delicately framed shot of a cat butt might have cost Matt the job, yet in this instance, it became the basis for a valued shared experience with the interviewer.

In these examples, no one is describing a wholesale violation of how to apply for a job—no one is sending in a poem instead of a resume. Instead, people are using the genres available and showing their competence at managing these genres at the same time that something distinctive, whether by accident or by design, slips through. The people evaluating the genre then have to make a call: Do they reject the applicant because of this mark of individuality (perhaps because it is poorly or inappropriately done)? Or do they continue looking further, because, after all, some evaluators would tell me that they wanted to hire someone with a sense of humor who could fit in at that particular workplace. In short, this points to the tension familiar to anyone trying to create a personal brand: You want to be able to show that you understand the genres and can produce something that can be compared to all other instances of its genre. At the same time, you want to be distinctive and also indicate that you are unlike all the others. That is not only a difficult line to walk, but it is also unpredictable how someone will evaluate your efforts when you try.

Workplaces also differ from each other in structural ways, and identifying those differences may require more information than a job description or a website will reveal. For example, I interviewed the head of staffing for a local hospital in the Bay Area. This hospital's primary staffing concern was having enough nurses to fill its positions, but its ability to staff was also shaped by its contract with the nurses' union. For legal reasons, the hospital is obligated to advertise every position broadly. But not every position will be filled by an external candidate. If a nurse within the hospital wants to move to a new department, because of the union contract, that nurse will have top priority. The position will be filled by an outside candidate only if no one in the hospital wants that position. At the same time, this HR manager felt that it was relatively difficult to dismiss nurses as well, also in part because of the ways in which the union would get involved in protecting its members. One of the consequences of this system is that while it is relatively easy to move within the hospital, it is more difficult to be hired into the hospital—and the hospital recruiters were far more cautious about screening candidates. Workplaces may be organized in ways that privilege some kinds of applicants and disadvantage others, and job applicants find themselves confronting an uneven field that will cause even the most beautifully formatted resume to be rejected for reasons the job applicant could never anticipate.

Hiring practices have resembled hourglasses for quite a while—for years applying for a job has involved creating the genre repertoire requested, a repertoire which standardizes how candidates represent themselves. And these forms are designed to make it as easy as possible to compare candidates, although for this to be easy, evaluators have to learn how to read the different genres with an eye for comparison. Once a job applicant has submitted the necessary forms or produced the interview answers, most of what the job applicant provides in the application is in turn evaluated by groups of people connected to different workplaces that evaluate and operate according to their own complex logics. Even applicant-tracking systems haven't significantly changed this part of the hiring process. The computerized systems simply encourage people to anticipate how that software will screen applications by adding keywords and applying other similar strategies to their resumes and cover letters. In general, for decades some parts of hiring have stayed the same, such as the pressure to use a standardized genre repertoire in order to be employable. What *has* changed, however, are the strategies people use. Now people present themselves as business solutions, as bundles of skills, experiences, and networked relationships, which affects the ways in which people create their genre repertoire.

Using standardized genres in the hiring process might seem like a handy way to compare candidates, but in practice it creates specific problems. From the job seekers' perspective, this standardization gets in the way of letting them present how they will in fact fit into a workplace. The standardized genre repertoire also forces employers to do quite a bit of work at interpreting what are essentially marketing documents to figure out whether they want to hire someone.

While I was doing this research, I asked Dave Dunning, a noted social psychologist, what psychologists know about how to find good employees. He answered that the best way to know if someone is going to be good at a task is to watch someone do that task. If you can't watch someone do the tasks that will make up their job, then the next best thing you can do is ask them to talk in detail about how they do the work. You don't want to ask vague questions like "How do you treat customers?" Instead you want to ask the most specific question you can in order see if you can get a vivid enough sense of how someone does a task from their answer.

For the most part, evaluating how well someone has done their job in the past is a far cry from evaluating how well they do the task of writing the resumes, cover letters, and application forms that people have to fill out in order to be hired. How many times will a workplace require that you summarize a complex life history in a resume format in the day-to-day tasks that make up a job? More generally, how often in a corporate job will you use the skills you have had to demonstrate in order to get the job—perhaps using a problem-solution-result format to answer a question someone asks you during your workday? There are very few examples in all the hiring processes I heard about in which people are showing how they would actually work in a job or are being evaluated for tasks that they would actually do. Instead, for the most part, I heard about instances of job candidates who were evaluated for their competence creating a genre repertoire geared toward making them comparable with and yet distinct enough from all other job candidates. What employers actually saw people doing as they explained why they should be hired typically involved a different set of skills than they will have to use on the job. True, what qualifies as competence has changed as the metaphor underlying employment has changed, but the basic problems with using these standardized forms remain.

So what are some of the ways in which the genre repertoire has changed? I mentioned earlier that resumes now are supposed to reflect the ways in which a job candidate can be seen as a bundle of business solutions for the company's problems. This means that the resume is supposed to compile evidence, often statistical, of ways in which the job candidate has solved similar problems in the past. Interview answers now are also supposed to be different. The problem-solution-result structure encourages speakers to present themselves as a business solution in human form, one that has provided other businesses with positive outcomes in the past. Finally, LinkedIn profiles are condensed glimpses into how a person might be a bundle of skills, experiences, and relationships—showing what someone has done in the past, listing their skills, and revealing a version of their professional network in the form of LinkedIn connections. While you still need to use a genre repertoire to get hired, how you are supposed to create this genre repertoire has changed. And how those hiring are supposed to interpret the repertoire has also changed; they are now supposed to see all candidates as bundles of temporary business solutions.

The solutions people are developing to deal with the problems of how standardized every job candidate has become are also inspired by this new model of the self. Networking now is often described as the way to get around both the ways applicant-tracking systems screen out applicants and the ways employers might not fully appreciate what a candidate has to offer from the too standardized application submitted.

Three   **Getting off the Screen and into Networks**

Job seekers and career counselors talk about networking as though it is a magical elixir—as though it is the most reliable path to getting a job. If the self is now metaphorically a bundle of skills, assets, qualities, experiences, and relationships—then relationships are the part of the bundle that people believe is most likely to get you a job. At the same time, it is the part of the bundle that career counselors think people are the most reluctant to enhance. Enhancing the self is not effortless. It requires discipline. It requires concerted and systematic attention. In job seeking, not every part of looking for a job is talked about as being a huge burden. However, people do tend to single out networking as the part that requires the most self-control to just keep doing.

These days, the perceived problem is rarely that people are reluctant applicants. Far from it. When career counselors run workshops, they often tell their audience that they spend far too much of their time applying for jobs online. Job seekers, they say, need to get offline and start meeting people in person. They need to exert some discipline over their natural impulses to stay home, and they need to stop imagining that they are being productive by spending hours upon hours submitting resumes. Instead, they should go out and network.

The advice that your social relationships will lead to a job is not new. Job-advice manuals in the early twentieth century suggested that your friends and relatives were a likely source for learning about your first or next job. The manuals might not have used the term *networking*, but people still quite reasonably thought that their social contacts could lead them to business contacts. Yet the reasons given for why people need to network have changed over time. This is in part because our media ecologies—the different types of media available for communicating with each other—have changed. Before, when people discussed how important the personal connection could be for finding a job, they were concerned about how difficult it was to learn about a job opening. In 1936, Charles Prosser wrote in his guide, *Getting a Job*: "Today it is more difficult than ever before to get information about occupations, to learn where employment is available, to get into contact with those having vacancies to fill, and to secure a position in the line of work you want to follow."[1]

Today finding out an opening exists isn't the problem. The internet has made it too easy to discover that a job exists and too easy to apply. As a result, advertised job openings often have too many applicants. To deal with all the applications, companies use automated mechanisms, applicant-tracking systems, to select the most likely candidates. When people talk about why networking is so essential, they focus on the problems created by job boards and applicant-tracking systems. Job seekers struggle to get recruiters, HR people, or hiring managers to notice their resumes. After all, regardless of how good of a fit you might be for a job, if no one ever looks at your application, you don't stand a chance. And the number of job applicants sometimes means that people only look through a certain number of applications. As one HR person explained, she knew hiring managers who only looked at resumes from certain sections of the alphabet. The people who screen applications don't necessarily look at the resume of everyone who applies. Of course, this isn't only a failure of a person faced with too large of a pile of resumes and not enough time. Getting past the applicant-tracking systems is the most common hurdle every applicant faces.

The resumes that the reviewer sees are presorted by an applicant-tracking system, if the company is large enough or flush enough. And the applicant-tracking system will often screen out appropriate and acceptable candidates through the vagaries of a keyword search, or because of some other sociotechnical failing. If the job description asks for someone with experience building international teams and the applicant mentions managing global teams instead, that person runs the risk of being rejected by a poorly programmed applicant-tracking system. Or the hiring manager may be willing to hire someone with four years of experience instead of five when faced with a promising enough candidate. But the applicant-tracking system will be programmed to eliminate anyone with less than five years of experience at a certain job. In the tech world, this can often become especially frustrating for both sides, for example, if the applicant-tracking system happens to be programmed to only select people who have three to five years of experience with a particular technology or programming language, but the technology or language was only developed two years ago. When catch-22s like these exist, getting hired involves finding some social route to circumvent the applicant-tracking system.

For all these reasons and more, job seekers and career counselors describe the online form of application as a black hole. So job seekers focus on finding ways to do an end run around the pile of resumes, and the principal way is by networking. Job seekers try to find an individual in the organization who will personally send their resume to the hiring manager. Ideally their helper on the inside walks over to the hiring manager's desk and places the resume in the hiring manager's hand. Throughout any discussion of job searching, there is a strong emphasis on the ways that face-to-face interactions are best. Having the company insider send the hiring manager a resume by email is clearly seen as second best, although my interviews lead me to suspect that this is probably the most common way that resumes actually circulate within companies.

As people explained to me these principles of networking, I was always struck by the potential risks to someone's reputation in referring people who might not be qualified or appropriate candidates. I was assuming that your reputation is in some way linked to the person that you choose to refer. Yet when I asked, people always hastened to tell me that passing along a resume did *not* mean that they were also actively recommending the person. They were simply passing along a resume, and they believed that this was widely understood. Some people would insist that they weren't recommending anyone, that they were only introducing. From the point of view of the people willing to circulate a resume, a networking connection was not a strong endorsement, only a strategy to avoid the black hole of job applications.

While some see other people as the ideal vehicle for having your job application noticed, sometimes people turn to older forms of communication because of their frustrations with email or applicant-tracking systems. People would tell me stories about how they used the manila envelope to their advantage, especially in moments when they had no access to someone within the company. One man in his midfifties was so excited to tell me about job-searching strategies he had just figured out in the previous month. All his strategies focused on tinkering with the different genres everyone uses to present themselves as hirable. He decided to transform his resume from one focusing on what he had accomplished to a resume describing what he hoped to do in the future, changing from achievement to potential. I understood this to mean that he was rewriting his resume to conform more to a self-as-business vision. And then, earnestly, he said that he had decided that mailing his resume to a hiring manager in a manila envelope was the way to go. He stressed how important it was to make a hiring manager physically open the envelope now that everyone in the business world is surrounded by email and online applications. Doing this, he thought, forces the manager to look at his resume, even if only because the manager is trying to figure out what on earth is actually in this piece of mail. Here the desperate attempts just to get noticed merge with people's beliefs about how older technologies require or inspire a different kind of attention.

Why the Need for Discipline?

New technologies of hiring have created problems for job seekers, and people's views of networking have been shaped both by the current media ecology and by changes in ideas about employment. Yes, earlier manuals on job seeking suggested practices similar to networking as the ways that people would prefer to find a job, but networking was by no means the magical elixir it is seen as today. In 1922, Norman Shidle in his manual, *Finding Your Job*, recommended turning to social connections only as a first step, because he saw it as the easiest and most problematic job-seeking strategy. He cautioned readers that they shouldn't be lazy, that finding jobs through social connections might lead people to taking dead-end jobs. By contrast, nowadays, creating social connections is described as an act of discipline, something that will feel like a burden.

As I mentioned earlier, relationships have become the part of the self that is seen as requiring the most disciplined management. Career counselors say over and over again that this is something you simply have to do, regardless of how unpleasant you might find it. One speaker at a packed workshop held in the city council chambers even compared it to dieting:

> I think the biggest faux pas people make is that they don't really realize that networking is truly work and it is something just like a diet. If you want to be mindful of what you eat, you've gotta do it every day. What invariably happens is that we are creatures of comfort. We have a long arduous search. We find something. It's the job that we were looking for. We're very excited. We get into our routine. And then we get tunnel vision. We get this sort of myopic focus that says, "Okay, this is my world now." And that was the old model of employment. That model does not exist anymore. One must be broad by nature now. And you always have to be networking.

When I first heard this comparison, it seemed odd to me. Why should meeting new people be equivalent to controlling how much and what kind of food you eat? What is it about networking—the most social part of being a business—that is so unpleasant? After listening to many workshop leaders explain that networking is crucial while acknowledging that it is a distasteful way of being social, I began to think that one of the main problems with networking is how much it encapsulates the changes implied with the new metaphor of self-as-business. Even your ability to be friendly and pleasant to talk to is now central to how you function as a business.

This "burden" of always being friendly can create a situation in which parts of this bundle you supposedly are as a business begin to clash with each other. Sometimes people have personal qualities that are supposedly part of who they fundamentally are as people and yet don't lend themselves to networking. Many people pointed out to me that if the primary way to get a job these days is through networking, then people who are introverted are at a disadvantage. People were quite unhappy that your personality now led directly to economic advantage or disadvantage. And they felt that the higher turnover in the job market meant that being an introvert was a bigger problem now than in previous decades. After all, nowadays you are supposed to be networking all the time, not just when you need a job. And people who described themselves as introverts also tended not to network while they had a job. Networking was seen as a task, and as a task it was something that had quite low priority for them while they were working. One job seeker explained to me: "So what happens is, when I get a job I usually put the blinders on. I do my job because in my own personal life, I'm married, got a wife, three kids, two grandchildren. And I've got a hobby, I like to raise chickens. . . . When I come home that's all I really want to focus on. I don't want to have to come home from work and then—oh, I've gotta go to this networking meeting and be doing that on a regular basis."

This cycle was something that career counselors were quite concerned that people, especially introverted people, would do—that people would enter into periods of expansive networking when they were looking for work and minimal networking when they had a job. Al, a volunteer career counselor, would often tell a story about a woman who systematically got 450 LinkedIn connections while she was looking for a job: "She ended up over a period of two and a half years, developing a network of 450 contacts. And when she started she had about five. She got the job, the network's gone. It's like vapor network. 'Cause she can't possibly stay in touch with 450 people especially when 25 percent of them are changing jobs every year. She can't even keep track of their emails.[2] When people come out of a job, they have to invest." Al was very clear that the network is a business asset, something you "invest" in, and should be investing in constantly. Counselors are concerned that people will only network during their job search, but it is something that today people are supposed to do all the time, not just as a signal to other people that they are looking for a job. Networking is supposed to be a routine and consistent (but also unpleasant) task for today's workers—hence the analogy to dieting. The metaphor of self-as-business shapes the ways in which people understand networking as simultaneously essential to job seeking and a problem—as a potential clash between how you like to manage your relationships and an external imperative to accumulate as many ties as possible.

While people have had relationships that are both professional and social for centuries, networking is a relatively recent way to describe this practice. The first use of the term to signify social interaction that I can find is in 1976: the British Institute of Management published a set of guidelines encouraging managers who had been laid off to get their next job by networking.[3] Even in this early usage, networking is about having strategic social interactions that are intentionally instrumental, but often ambiguously so. Networking mixes the personal and the professional, just a little, and in complicated ways. People you meet in a networking event could potentially turn into your friends, as well as perhaps be people who might be useful for you to know professionally. Conversely, you can network with friends, but going to dinner with your friends is not a networking event. Networking is about potential, about collecting as many relationships that could be activated for your interest but are not always immediately activated. Often it is ambiguous whether a connection you started through networking will ever pay off, but you are supposed to keep networking and keep hoping.

Informational Interviewing

Even the genre that seemed like the most instrumental form of networking—an informational interview—is supposed to contain a modicum of ambiguity. Richard Bolles coined the term for this kind of conversation in the 1982 edition of *What Color Is Your Parachute?* after calling it a "research interview" in the editions published in the 1970s.[4] The informational interview is one of the genres that job seekers are strongly encouraged to be competent at in the United States. In an informational interview, you discuss with someone, over coffee or the phone, aspects of the company they work at or the job they do. Having coffee or chatting on the phone doesn't seem like something that would require a lot of instruction. Yet informational interviews are seen as so emotionally charged and challenging that many organizations for job seekers dedicate entire workshops to explaining how to do them.[5] And those leading the workshops are quite clear—these interviews are overtly instrumental. There is supposed to be a one-way exchange of information that will be useful for the job seeker in orienting him or her toward a job's demands or a company's organization.

Yet even here, in a conversation that is supposed to be about employment, there are limits to how instrumental the job seeker is allowed to be. When career counselors describe how to manage this genre of social interaction in workshops, they always hasten to caution job seekers that they should not explicitly ask for a job during these conversations. Indeed, they should mark their competence at this genre by creating a certain level of ambiguity about what they want the person to do for them, even though simply asking for an informational interview is clearly an instrumental request. One career counselor recommended that her clients avoid using the words *informational interview* when they request a conversation with someone, unless they recently graduated from high school or college.

Recent graduates, she believes, are given far more leeway as beginners in the job market, and they won't make anyone uncomfortable by suggesting this type of conversation. For everyone else, she recommended that they represent their request in terms of research and claim that they are mulling over what new directions they might take in the future in terms of their career. And here, she stressed, it is important that the emphasis be on your career instead of a job. So this career counselor suggested as an opening to frame a request for a meeting sentences like "I am in an interesting place right now, and I am researching where I can best contribute and make the most impact."

When counselors try to explain that even informational interviews should have an element of ambiguity, they often focus on what a job seeker is supposed to do with his or her resume at these interviews. Career counselors stress that a job seeker is never supposed to volunteer his or her resume during an informational interview. The person granting the interview is supposed to be in control over when and how the resume circulates. Some career counselors encourage people to bring their resumes along to the interview in case the person granting the interview requests the resume. Others, however, suggest that even having the resume present may be too much temptation to an anxious job seeker. These counselors recommend that the job seeker leave the resume at home. If the person the job seeker is meeting happens to ask for the resume, then the person should promise to email it as a follow-up. Not having a resume becomes a good excuse to keep interacting after the informational interview.

Some job seekers I interviewed found this need for ambiguity even in an informational interview a bit baffling. Given that the whole point of the conversation is to get a job, and the conversation wouldn't be happening in the first place if the job seeker wasn't looking, why disguise this? Yet career counselors insist that a certain degree of ambiguity is essential for this interaction to work smoothly. While everyone might be willing to acknowledge how instrumental the conversation is when asked, in the moment of the interaction, the model requires that there be enough ambiguity that the communication can also seem to be just about establishing a connection with someone else. Informational interviews, like all other networking interactions, are about being both instrumental and friendly, and the person granting the interview is generally much freer to be explicit about the instrumental nature of the interaction than the person asking for the interview. The dilemma that informational interviews always pose for job seekers is how to maintain this ambiguity between the instrumental and the friendly throughout the interaction.

Paying It Forward

When career counselors discuss networking, they often suggest that it can be a moral activity. They realize that at first glance, this seems implausible to their audience. After all, networking is so commonly described as overtly instrumental social interaction disguised as friendliness that their clients will express their reluctance to network by insisting that it feels too phony and too contrived. How can this be moral? In workshops and one-on-one counseling sessions, career counselors try to reframe networking as a moment to imagine how the potential networker can help a stranger—a pay-it-forward moral commitment. Networking is a chance to establish a relationship by helping someone else, and at any moment that you are networking, the focus, they say, should be trying to figure out if you have any information or connection that might be useful to the person you are talking to.

In insisting that networking is a moment in which people can and should concentrate on paying it forward, counselors are reframing networking as an exchange relationship, a relationship in which you focus on the information and connections to other people that you can provide, instead of what you can receive. Yes, it might be instrumental, but it should not be done in a self-absorbed form. It is pointedly *not* a barter relationship or a market relationship, in which there is an immediate quid pro quo—in which one person offers information and in the same conversation the other person offers information he or she imagines is of equal value. Even in the common phrase *pay it forward*, people are recommending that everyone think about these exchanges as taking place over extended periods of time.

People are supposed to be establishing the kinds of exchange relationships that anthropologists have long documented in places or moments that are not dominated by the market. These are relationships in which you give to other people in order to establish a durable connection. In many market relationships, you have an immediate exchange—you give someone a certain amount of money and you get an object or a service in return. But in nonmarket exchanges, you give in the hopes of eventual reciprocity, although the reciprocity is never guaranteed. You drive a friend to the bus station, and hope the next time you need a ride to the bus station, your friend will give you one. But they might not be able to. And the fact that they can't that day doesn't destroy the friendship (well, not normally). You don't calculate precisely how many favors you have done others or expect them to reciprocate exactly.

Indeed, in the noncapitalist exchange systems that I have studied as an anthropologist (mainly in the Pacific), part of the pleasure and anxiety about being a part of these exchanges is in never knowing when and how someone will reciprocate. Often, exchanges take place between families, not individuals. For example, you might think that the Tua family should bring a certain amount of food to your family's feast, because, after all, your family provided quite a bit of food the last time they had a feast. But you don't know if this will definitely happen, and if so, how much they will in fact contribute. And the amount that family contributes will be an indication to you of how the Tua family values the relationship between the two families. These exchange relationships have historical depth and multiple ways of being interconnected. The debt incurred by these connections is never fully obliterated—your family will in turn be expected to bring food to the Tua family's feast the next time they have one.

I am evoking in broad strokes a different exchange relationship to try to give a concrete example of how exchange relationships could be supposed to happen when you don't hand over money for goods or services in that very moment. This other kind of exchange takes place over long periods of time, and what exactly will be exchanged is nebulous. In the case I described above, it will probably be food—but what kind of food? If the Tua family brings corned beef to one feast, your family is not obligated to bring corned beef to their next feast. And if your family carefully measured the corned beef, and brought exactly the same amount of macaroni salad to their feast as the corned beef they brought two months ago, this would seem a bit strange, a bit too calculating. Indeed, your family might bring two dishes instead of one in return—not to metaphorically pay interest but to express both your family's generosity and how much your family values the historical depth to the relationship between the families.

Paying it forward similarly involves a kind of reciprocity that is neither immediate nor precisely equivalent. But there are ways in which networking as paying it forward doesn't resemble these other kinds of exchange relationships. For starters, networking does not often involve families—it is a relationship of exchange between individuals, first and foremost. A brother or cousin may in fact be the connection that is offered, but this does not mean that someone's entire family has just entered into an exchange relationship with another family. And, sometimes, when you are offering information to someone, it is ambiguous whether you are hoping to establish a reciprocal relationship with that person down the line or hoping to establish a reciprocal relationship with a nebulous and ill-defined community. If you contribute to a community in which people are constantly helping each other, perhaps eventually someone will help you when you most need it. Hoping to be part of such a community is a common reason people give when they explain why they do informational interviews. Yet unlike other exchange relationships, it is never clear who is part of this community and who isn't. You are supposed to network with strangers, turning them into acquaintances, and to try to build an ever-expanding set of connections without any concern over what the boundaries might be. The point is to be as inclusive as possible, although in practice there are all sorts of subtle ways in which certain types of people are excluded.

Weak Ties, Strong Ties

I have been talking about some of the complicated reasons why people might not like networking: it is about nourishing ambiguous relationships that are both instrumental and personal in ways that make Americans uncomfortable; it is about being expansively social when someone might prefer to be introverted; it is about requesting favors from people when you haven't established a sufficient history of exchange with them. It is, in short, socializing with the aim to accumulate as many weak ties as possible and make these weak ties into connections that will hopefully lead to a job.

But why weak ties? In part, this is because of how influential Mark Granovetter's sociological study, *Getting a Job*, has become—a study in which he demonstrated how effective weak ties were for white-collar men transitioning from one job to another. In the early 1970s, Granovetter researched how people found out about jobs, determining after interviewing one hundred men that most people found out about possible new jobs through weak ties, through people whom they only vaguely knew. After all, as Granovetter pointed out, a person's close circle of friends and family are as likely to know about a job's existence as that person is. It is the more distant network connections that might have valuable new information with regard to possible jobs. In the very first workshop for job seekers that I attended, the workshop instructor explained that research has shown that people do not get a job from their friends, but rather from the friends of their friends—she knew about Granovetter's findings. And people kept mentioning the strength of weak ties in other workshops and interviews. In short, networking has become, for the most part, the conscious attempt to put Granovetter's sociological insights into concrete practice to your best advantage.

LinkedIn has helped make this easier. One career counselor

explained to me that LinkedIn has been invaluable for helping her explain the concept of weak links. Before LinkedIn, the job seekers she counseled had trouble understanding what it might mean to seek out the friends of a friend. Now, all she has to do is explain that they are looking for second-level connections, and immediately her clients understand, largely because LinkedIn classifies every person in terms of the level of connection that person might have to you (first, second, or third). If you have a second-level connection with Brad on LinkedIn, it means that you are connected to someone who also is connected to Brad on LinkedIn. This isn't exactly what Granovetter meant by weak ties; in his examples, the two people in question had met each other but didn't know each other well or see each other often. But it is a close enough analogy. As more people use LinkedIn, it becomes easier for people to understand that when they are looking for a weak tie, what they are doing in practice is searching on LinkedIn for second-degree connections in the company they are hoping to join.

But all people know from their LinkedIn profile is that these are second-degree connections. They have no sense of what kind of link this is, that is, when people first met, how much contact they have with each other, or how much they actually like and respect each other. Networking nowadays might be about accumulating connections to other people, but it involves collecting connections indiscriminately. Just the fact that the connection has been made is enough. People are often collecting contacts—business cards and LinkedIn connections—focusing on the number of contacts instead of the quality. This can be a topic when people talk about networking, as they discuss how meaningful the conversations are supposed to be when establishing these connections. In the discussions I heard, people focused on the initial contact, not on how people develop or build long-lasting complex relationships over time. In the rare moments in which they talked about sustaining these weak ties, they recommended sending a link to a news article or blog post that might interest the person out of the blue. These are, after all, weak links.

Why do people like the idea of weak ties so much? What is it about Granovetter's model that has captured the imagination of career counselors and job seekers? In part, as I mentioned earlier, weak ties provide a solution for a perceived dilemma that new online job boards and application-tracking systems have created. But turning to weak ties is also a response to a decades-old problem for the job seeker—from the outside, companies often appear to be black boxes. Job seekers understand that there may be complex hierarchies within a company, and that it is often unclear who has the power to offer a job. What exactly the job involves can also be murky. One of the longest-established career centers in Silicon Valley used to have files upon files of clipped information about companies so that job seekers could do research on who exactly was in a company and how exactly it was organized. Now, some of this information is available through diligent internet searching, although most people seem to focus all their energy on figuring out this information through LinkedIn. Even with all these avenues, the information is often difficult enough to find that people turn to connections within companies in the hopes that company insiders can shed light on how companies actually function.

At the same time, when job seekers use the network as a metaphor to imagine a company's internal organization, they are not imagining themselves at the center of the network. This can be disheartening, especially since all the social media sites people use regularly these days encourage users to imagine themselves at the center of their own network. The view of your network in LinkedIn or Facebook begins from your own profile. Yet for the job seeker, the hiring manager is at the center of the network. The question the applicant wrestles with is, what social paths can the job seeker follow to find a way to connect with this center? And as job seekers often find themselves interviewing at a number of places in a month, they are constantly being reminded that others are the centers of networks, and they are displaced. Collecting weak links is a way to think that there is something concrete you can do to get to the center of the network.

The challenge for job seekers has switched from Granovetter's moment, when you needed weak links to know about the job, to now, when people believe that you need weak links to be known. Now that Granovetter's work is the everyday common sense of career counselors and the hard-won wisdom of job seekers, networking has become a very conscious and sometimes burdensome form of expressing agency. People are constantly told to network, to transform even the most casual conversation with a newly met person into an opportunity to circumvent the obstacles technologies have created as people try to connect to jobs. Thus networking is always hopeful. But as someone fails repeatedly to get a job, the network also becomes a sign of their inability to elicit proper social support from their community, however vaguely defined that community might be. Now that people know what weak links are supposed to do, these weak links are also regular reminders of weak selves.

This, I want to point out, is an unusual moment in Americans' relationships to technologies. Americans often want technological solutions for what they perceive as technological problems. Indeed, they often want technological solutions for what they view as social problems as well. This is one of the infrequent moments in which Americans want a social solution—networking—for what they perceive as a technological problem.

Workplace Ties

So much has changed since Mark Granovetter's initial study that I was curious if weak ties are still the most useful kind of social connection you can have in finding a job. I've come to the conclusion, however, that thinking about networking ties in terms of whether they are weak or strong is the wrong question to ask in the first place. Instead, what seems to matter most is whether the person recommending you, whoever he or she may be, knows what you are like as a worker. How did I come to this conclusion?

To see if Granovetter's findings are relevant nowadays, I needed a sample of white-collar workers explaining how they found their current job. I was fortunate, because, as I described in the introduction, an organization geared toward helping professionals had weekly meetings in which anyone who has gotten a job stands up at the start of the meeting and tells their "success story."[6] This organization had been filming these success stories since January 2012, and it allowed me to have access to the videos. I watched 380 stories in which white-collar workers explained what job-searching strategies led to the job they just got. Readers who are interested in finding a job themselves should keep in mind that what I am describing is not a random sample; it is a descriptive sample. Everyone was preselected. They were unemployed professionals in the Bay Area who had been allowed to participate in this organization. The group composition was as follows: 53 percent were women and 47 percent were men; 89 percent were over 45 years of age; for nearly half the group (44.6 percent), a bachelor's degree was the highest degree obtained; and 19 percent had an MBA and 22.6 percent had a master's degree. If you are wondering if their experiences have any bearing on your own, please keep in mind how close you are to this sample matters.

In Granovetter's study, social connections were useful for finding out about the job; 26 percent of the people found out about jobs through formal means, and the rest found out through word of mouth.[7] In my study, social connections still were useful, although perhaps not as useful as in the early 1970s. In my sample, people found out about jobs through word of mouth 37.5 percent of the time. Almost as frequently, 35 percent of the time, they found out about jobs because they were contacted by recruiters or staffing agencies (often because of their online activity). They discovered that there was a job opening online 26 percent of the time.

Networking was useful for many, but not everyone; 45 percent of people said that networking had helped them get the job. I wanted to know what kind of social connection had helped, so I classified the connections using the categories I had frequently heard mentioned in the interviews I did and the workshops I attended: friend/relative (a strong connection); friend of a friend/acquaintance (a weak connection); informational interview; networking event; volunteering; former coworker or boss, and so on (see fig. 1). I also used a category

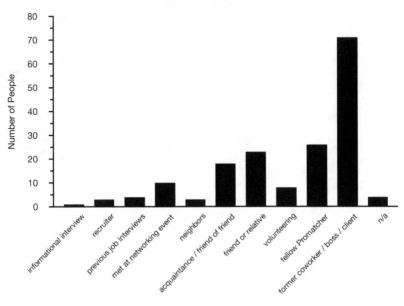

TYPES OF PERSONAL CONNECTION

Y-axis: Number of People (0 to 80)

X-axis labels: informational interview, recruiter, previous job interviews, met at networking event, neighbors, acquaintance / friend of friend, friend or relative, volunteering, fellow Promatcher, former coworker / boss / client, n/a

Type of Connection

Figure 1. In my sample of white-collar workers who had found jobs, 171 people said personal connections were useful. This chart shows the types of personal connections people reported.

that was specific to this particular group—fellow job seekers met at the organization. This was an organization that was largely run by the job seekers themselves. They organized and taught all the workshops and ran the meetings. This gave job seekers a chance to work with each other, to learn who was reliable and who was flaky. So I was struck when I saw that while 23 people had been helped by a friend or relative, 26 people had been helped by a fellow job seeker at this organization. Both categories are overshadowed by the number of people helped by someone they used to work with at a paying job— 71 people. But if you don't focus on the difference between weak ties and strong ties, and instead think about whom people worked with and whom they didn't, the numbers are striking: 61 percent mentioned being helped by someone who could speak in detail about what they were like as a worker, 17 percent were helped by people

THE STRENGTH OF WORKPLACE TIES

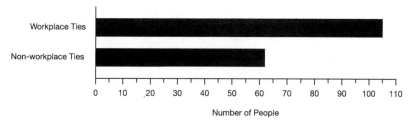

Figure 2. This chart shows the results of categorizing the personal connections in my sample in terms of whether people worked with the job applicant or not.

whom they didn't know very well (an acquaintance or someone they met at a networking event), and 13 percent were helped by a friend or relative. In short, for the people in my descriptive sample, it seemed to matter more if your networking connection could, with authority, tell others about what kind of worker you had already been (see fig. 2).

This finding makes a considerable amount of sense if you reflect on what has changed in the hiring process since Granovetter did his research. In the early 1970s, people looked for jobs in a media ecology composed of newspapers, postal mail, telephones, and face-to-face interaction. People might hear of job leads in supermarket parking lots or weddings, when they ran into someone they hadn't seen for a long time. In short, when people first began openly talking about networking for a job, networking was a strategy for learning that a job existed in the first place, not, as it primarily is today, a strategy for calling a hiring manager's attention to a resume.

In today's media ecology, the difficulty lies not in having the potential employee find out about the job. The difficulty lies in distinguishing your online job application in a pile of resumes. Our contemporary media ecology has fundamentally changed the hiring process from Granovetter's day, so much so that new definitions and practices exist for what counts as a weak tie, and how you activate a weak tie, as well as why you might require a weak tie in the first place.

So what are some of the differences between job searching in early

1970s America and job searching now? As I discussed, there is a different media ecology, and as a result, the problems in job searching have changed. People no longer struggle simply to learn that a job exists—online job ads make it much easier for people to find out about jobs. Instead, they now struggle to have the right people in the company notice their application. In addition to changes in hiring technologies, the average number of years a worker stays at the same job has dropped. By the time Granovetter was doing his research, companies had spent decades trying to present the loyalty and security of being a company man in a positive light. This is no longer what companies or even many employees expect. Even government-endorsed work blogs have slogans such as "Career-realism—because every job is temporary." In my sample, people had an average of four jobs from 2004 to 2014. As a result, job transitions have shifted away from being the more unusual events they were in the 1970s—indeed, Granovetter had to find his interview subjects through carefully scrutinizing five years' worth of telephone books. Now transitions are not only the norm, but moving from job to job is widely seen as essential for crafting a successful career. Furthermore, in the 1970s, Granovetter found that people's weak links were to people they had known for at least five years, if not longer.[8] Today, the weak links people described in my fieldwork are often virtual connections that people actively pursue by starting conversations on LinkedIn or retweeting a potential hiring manager's tweets. Lastly, while you might want to be ambiguous about how instrumental you are trying to be while networking, it isn't a forbidden or problematic reason for talking to people. By contrast, in Granovetter's study, openly acknowledging the instrumental motivations behind an interaction would, as he put it, "curtail the relationship."[9]

You might imagine that workplace ties wouldn't be all that effective, because why would people at a company help their competitors hire good people? This is another moment in which having an employee view him- or herself as a business might clash with the best interests of the company. Helping coworkers or former coworkers get jobs at different companies paves the way for you to get another job. Building networks of people across a range of companies, even those your employer is competing with at the moment, means that you might have options in the future when your interests and your company's interests diverge.

Networking: Luck versus Risk

Weak ties, strong ties, workplace ties—these are all ways of classifying social connections that don't intrinsically reflect on the change in metaphors that I am arguing has so affected hiring over the past forty years. Yet much of the advice that job seekers often hear about how to accumulate weak ties is fundamentally linked to changes in people's understanding of the employment contract. I found two different approaches for networking: networking based on luck and networking to manage risk. Both are based on what it means to view yourself as a business. They are simply two different interpretations of how best to be a business. Networking based on luck involves collecting as many weak ties as possible, and this is a method for dealing with the uncertainty that you find in any market. The other approach revolves around minimizing the risk involved in being a business for hire; you are supposed to do targeted networking to try to find the best alliance possible. And it turns out that you can't do both at the same time—maximizing luck involves a different set of strategies than minimizing risk.

I only realized that there were two distinct approaches while chatting with Al, a cofounder of a church-based organization focused on helping the unemployed. This organization tries to create opportunities for job seekers to network with each other, under the assumption that even though the person they are talking to may be unemployed, that person knows a range of people who have jobs and could potentially help. Yet job seekers are unlikely to come to an event whose sole purpose is networking, and especially not networking with unemployed strangers. They need a reason to go to a particular meeting, so the organization invites motivational speakers. After a talk from a motivational speaker, Al and I went out to lunch at a sushi restaurant. I asked him what he thought of the talk, expecting him to say vague positive compliments because he is such an optimistic and generous guy. In a rare moment of critique, Al admitted that he thought the speaker was promoting the wrong approach to networking. The speaker had insisted that people never know where the contact to their next job might come from. She said that everyone should talk to the people in their social world that have contact with people from all sorts of walks of life—your mechanic or the teacher of your child's kindergarten class. She had vivid stories about people who found jobs through everyday encounters: Someone struck up just the right conversation with the person squeezing peppers next to them in the supermarket. Or a man chatted with his haircutter about having just lost his job as a CFO, and the haircutter had the week previously cut the hair of someone who was complaining about how difficult it was to find a qualified CFO. Her description of networking was all too familiar to me by that point. I had heard many career counselors suggest that people find jobs serendipitously because someone they happen to know happens to know someone else who knows a hiring manager or recruiter.

Al had two main objections to this approach to networking, and he strongly advocated a more targeted approach. First, he felt that this strategy relied too much on luck, and hence took an indeterminately long time to be effective. This was a waste of resources— wasting people's time and money alike. He felt that there had to be better strategies that would cut short the time spent searching for a job. Second, he believed that this strategy often meant that people were not consciously finding good jobs. They were not only relying on luck to end their job search, but they were also relying on luck to end up with a job at a company that was thriving. In Silicon Valley, so many companies are temporary ventures and might go under after six months or a year. Al believed that if people simply had a different approach to networking, job searches would work out differently.

This other approach to job seeking takes seriously the idea that being hired is metaphorically an alliance between two businesses. You should do research on the business that you are about to ally yourself with, just as businesses do due diligence before entering into any form of partnership. Thus the focus of networking should be different. Rather than hoping to find a lead to a job, you should choose five to eight companies as your target, although the numbers recommended can vary wildly. I have heard counselors recommend anywhere from five to fifty target companies. You then research these companies to determine whether they are viable companies. Al strongly recommended paying attention to which venture capitalist was investing in the start-up, if it was a start-up company. He believed that a handful of venture capitalists were wildly successful because they did such a thorough job of analyzing potential investments. If one of the top five venture capitalist firms was willing to invest in a company, it probably would be a good enough risk for a job seeker. Of course, this means using a venture capitalist firm's standards for determining a good business risk as the proxy for doing your own research into a company.

What Al was outlining were two different approaches to networking. He was criticizing an approach that asks job seekers to focus on accumulating as many alliances as possible to address the apparently random fashion with which information about hiring circulates nowadays—to manage uncertainty. Al instead endorsed an approach expecting job seekers to anticipate the dangers of entering into an alliance with a business that might fail—to manage risk.

In practice, these two different modes of networking involve different sets of online strategies as well as different sets of offline strategies. If you believe that the goal of networking is to manage uncertainty, your efforts are all geared toward creating as many alliances as possible. The more people you know, the better. After all, anyone might happen to be the bundle of relationships that can, if approached properly, initiate the interactions that will lead to a job for you. The more business cards exchanged, the better. The more LinkedIn connections you have, the better. Counselors subscribing to this model will recommend that people join LinkedIn groups— topic-specific forums in which people circulate news articles and tips for how to deal with certain job-related dilemmas. These virtual conversations are also supposed to lead fluidly to more LinkedIn connections.[10]

People advocating this approach to networking also suggest that you meet people in similar offline circumstances as well. The organizations I visited strongly encourage people to network with each other at the job seeker meetings—indeed every event has a thirty- to sixty-minute introductory period set aside for networking, and an hour after the event is also set aside for networking. Even after hearing this countless times, I still find it somewhat unnerving to hear "Okay, now start networking" loudly announced by an organizer as the group's signal to become more openly instrumental about their interactions. While some job seekers occasionally express some bewilderment that they are being encouraged to network with other unemployed people, their reservations are explicitly counter to the approach of networking being endorsed. Counselors will tell people repeatedly that you never know where the next job lead may come from—just because someone is currently unemployed doesn't mean that they don't have many friends and relatives who are employed and can be of potential benefit.

Occasionally the organizers of these events will seek to demonstrate that you never know when you will meet someone useful for your job prospects by leading an in-person activity that mimics what you are also supposed to do with your LinkedIn contacts. They will announce that they are going to connect seekers with potential company insiders and then ask if anyone is searching for a lead in a particular company. Someone will announce that they want to be connected to a person at Apple or Genentech, and then the crowd of fifty to one hundred people are asked if there is anyone who will admit (publicly) to knowing someone at that company.

This is asking simply for a connection, a tie without context or history. People realize that this form of networking has some obvious problems. Most of the companies mentioned in these moments are very large. Just because someone is employed at the company doesn't mean that they know the hiring manager for a certain job. The people asking for a connection also realize that all they are really asking for is a way to get a resume to a hiring manager that circumvents the applicant-tracking system. They understand that the person can't vouch that they will be a good worker.

When people network with those with whom they have had only the most glancing of conversations, it becomes more difficult to transform these connections into relationships that will cause someone to act. People may not be able to mobilize their networking connection even as a way to be introduced to someone else by email or LinkedIn, let alone to circulate a resume. One job seeker explained to me: "There was one job that I applied for and my inside contact, she told me that, 'Oh this job is filled.' And then I get another email from the HR saying, 'Are you still interested in this position?' What the heck? So you never know. Sometimes your inside contacts, you don't know if they're really helping you or they're just telling you they're helping you." When you request a favor under these circumstances, you have to hope that people will be willing to act out of sheer goodwill. After all, in these circumstances, there is often no reason why they would feel any obligation. Instead, you are often counting on a more elusive sense of connection to an ill-defined community of job seekers and soon-to-be-job-seekers. You have to hope that the person you ask wants to pay it forward. While this notion of networking depends on the idea that weak ties are the most promising ties, in practice, referrals or other help finding a job is often uncertain, since these kinds of weak connections are so unreliable.

Networking to manage risk, by contrast, involves a different set of online and offline practices. Online, you not only research the likely companies, but you also target potential allies or advocates within the company on Twitter, LinkedIn, and Facebook. People advocating for this perspective recommend that job seekers spend five to eight months establishing relationships through virtual forums with members of the desired companies. They suggest striking up casual conversations in LinkedIn groups, or following company employees on Twitter and then retweeting their tweets. After a while, you are supposed to send web links to items that might interest these people given what you have already gleaned about their preferences from previous online interactions. According to advocates of this strategy, virtual attention will get you everywhere, but it requires a stretch of time before it pays off. The downside is that you can not do this in response to a particular advertised job description. Instead, you do this in order to have people in place as your advocates when a job finally opens in that company. At the same time, once you are in a job, you should continue maintaining your contacts as much as possible—periodically checking in with people just in case. Both forms of networking share in common the strategy that you should be as instrumental as possible about simply building a reliable channel in the first place, but one form of networking is far more dispersed, and the other is far more targeted.

In the United States, these two different models of networking have advocates from different class positions. Those who recommend serendipitous networking, for the most part, are recommending this to midlevel managers, to white-collar workers who in the Bay Area are looking for jobs as product managers, supply chain managers, computer programmers, and biotech engineers. Those who recommend strategic networking that focuses on risk are imagining executives in corporate America. Both groups of job seekers will report that they are looking for a job, and often any job that they can do well, but midlevel professionals are encouraged to view the world of work in terms of uncertain and accidental connections, and executives are encouraged to see the world of work as business risks that they must navigate in as savvy a way as possible.

This has consequences for how people then locate responsibility when the jobs don't materialize. The job seekers who view networks in terms of luck are far more likely to uneasily blame their network—both their friends and their chance encounters—for failing to be kind. The job seekers who see networking in terms of risk are far more likely to blame themselves for not assessing the risks effectively enough or being effective enough to make a job materialize.

While I have been describing networking with risk in mind as a different approach to networking than networking with luck in mind, in practice even those who advocate that networking should revolve around evaluating potentially risky business alliances still believe that lots of luck is involved in networking. But they believe some control is possible. They recommend choosing target companies and strongly urge caution about which business you choose to align yourself with. They insist that job seekers should not make choices about jobs from a standpoint of desperation. Ideally, you should not think frantically: "I need a job, any job, and any job is better than no job." They fully realize that this is a privileged position, and sometimes people do in fact need to take a job, any job, because they need an income. Similarly, they also say that you never know where the next useful connection will come from. Networking for luck is almost always a theme that emerges as people discuss the practicalities of networking, regardless of the dominant model they subscribe to. No matter the approach, people talk about luck, but the reverse doesn't often happen. Those who advocate networking based on luck don't also talk about doing a lot of research on a company or assessing the risks involved in taking a certain job.

Yet when people are evaluating job candidates, luck rarely gets mentioned. When recruiters and hiring managers are seeing the traces of this networking luck, that is, someone's job history, they often interpret it as strategic moves toward fashioning a career. When people read resumes, they view former companies as examples of people's judgment about how to manage the risks of aligning themselves with a particular company. A career has become a pastiche of jobs, and contrary to many people's actual experiences, the pastiche is often imagined to be under the worker's control. According to the logic of personal branding, all the jobs are connected to each other because of how the applicant's personal qualities led him or her to make consistent decisions. When I talked to Kevin, a hiring manager, about how an increasingly rapid job turnover was affecting people's working lives, he described the risks that people were taking with each new job. He gave an example of how people need to make sure that they look reliable and strategic enough on their resume,

giving a detailed example of how one person might think through a
transition from company to company:

> Someone maybe worked at Clorox and they were in the management
> track. Clorox has a management track, and they spent ten years in the
> management track and worked their way up to director or whatever.
> Then they went to a technology company. They thought: "Hey, everyone
> is making a killing in technology, maybe I'll do that." And then, well,
> technology—not everyone is making a killing, some people are making
> a killing, a lot of people are just getting by. And so perhaps that didn't
> materialize to the extent that they thought it would. So then their next
> job is critically important to whether they get back. They have to pick
> the next company very well. Or get lucky—a lot of it is luck. So they
> come to another company and then they decide that company is not the
> right company for them. They just did the technology company, and it
> wasn't the right company for them, *and* a quick move. So now they have
> had a solid company for many years and then the technology company
> for a few years, and they go to another company, and they only stay there
> for one year because they were realizing it wasn't the right company for
> them. So now, for their resume, this next one they have to go to and stay
> there for a while. Even if it is not the ideal fit for them.

Regardless of whether the job seekers themselves experience the jobs
they have gotten in terms of luck, the recruiters and hiring managers
will often read people's work history as a glimpse into the kind of
career strategies they make, attributing much more conscious choice
to what job seekers often experience as random happenstance.

Your network is crucial in new ways for managing your career trajectory as people are expected to spend less and less time at a single
company. People are consciously managing their networks, in part
because they are constantly being told to network, and in part
because many of the contemporary technologies people use to communicate with each other make these networks visible as a matter
of course. As a consequence, jobs have become valuable not just as

places of employment, but as places where you can meet people who will help you get jobs in the future. As one person told me about how to devise a career trajectory these days, "It is the relationships, not the job." Supposedly your career no longer depends on the job you have, but the people you stay in touch with.

Yet put into practice, this emphasis on networking can create a conundrum for people trying to understand how hiring works as a system. If the current dominant metaphor for hiring has become that we own ourselves as though we are a business, an embodied bundle of skills, qualities, experiences, assets, and relationships, this emphasis on hiring through networking can make it seem as though selecting people based on their relationships is in conflict with selecting people based on their skills. This is a recurring frustration for people, that hires seem to be based more on whom you know rather than what you know. If people are seen as a business, networking privileges one part of this bundle of attributes over others, making an emphasis on skills less likely.

There are a number of structural conditions that can make networking a challenge; some of the conditions the people I interviewed mentioned quite openly, and other conditions no one ever discussed with me. People would talk to me about the ways in which their company was organized so that they were unable to build relationships at work that could potentially lead to valuable connections afterward. One job seeker talked about how, as an administrative assistant, she was largely assisting her boss, scheduling his appointments and taking care of the paperwork for his various projects. She simply wasn't located in a space in her office where she could have casual conversations with her coworkers. Even in those companies that publicly pride themselves in creating as many opportunities for interaction as possible through spatial layouts and work-related social activities, some employees still felt isolated because of the ways their work was organized.

Others would point out to me that the nature of work was such that people did not have time for networking—be it on their own behalf or by giving informational interviews. People can work long hours in the United States—according to a Gallup poll, in 2014 the average work week was forty-seven hours, and "nearly four in ten work at least 50 hours."[11] In these contexts, people found it difficult to find time to work, have a family, *and* devote time to networking. These pressures were one of the reasons that people often talked about networking in terms of self-discipline. Granting informational interviews was often complicated in situations of widespread time shortage, and job seekers would comment on how difficult it was to get people to agree to an informational interview.

These were the concerns with networking that people openly discussed with me. There were other bigger-picture issues that no one seemed to mention. For example, sociologists have long been critical of relying on networking in the hiring process. Certain people are often excluded from hiring possibilities when hiring revolves around networking—people from the wrong class background, the wrong ethnic background, the wrong gender, the wrong religious background all may be unable to have the social connections that will allow them to get hired into the jobs they want. As Granovetter points out: "If those presently employed in a given industry or firm have no black friends, no blacks will enter those settings through personal contacts."[12] This remains the case. Laura Rivera's book *Pedigree* reveals how elite networks and marks of an elite background are crucial in how employers select who gets hired as management consultants, lawyers, and similar jobs. In her words, "Hiring decisions that appear on the surface to be based only on individual merit are subtly yet powerfully shaped by applicants' socioeconomic backgrounds."[13]

This downside to networking, however, was never discussed. I found this odd, since the form of discrimination that people talked about frequently and openly was age discrimination—a discrimination that might easily lend itself to a discussion of how networking can fail people. After all, there aren't all that many instances of cross-generational social activities that people discussed doing with me. The social divisions between generations in the Bay Area can be stark.[14] Yet in discussing how pervasive age discrimination seemed to be, few people whom I spoke with saw this as one of the outcomes of the ways hiring was increasingly revolving around networking. No one was commenting on how when an old boy's network becomes a young boy's network, there can be new categories of people left out in the cold.

Four　　　　　**Didn't We Meet on LinkedIn?**

When I was exploring if hiring would be a good research topic, I had a long interview with Tiffany about all she did on LinkedIn as she prepared to graduate from Indiana University. I asked her at the end of the interview if she had any questions for me. "Yes," she said, "how do I use LinkedIn?" I was so confused: "But you just told me how you use LinkedIn." "I know, but how do you *really* use it?" This may have been my first interview that ended with "What is LinkedIn good for?" but I soon got used to being asked this question. The older job seekers I interviewed in the Bay Area were often equally uncertain about LinkedIn. Many talked about how a LinkedIn profile was essential for a job search. LinkedIn's importance was widely acknowledged. But the details about why LinkedIn was in fact so important were often a bit vague, and how to use LinkedIn effectively was often an open question.

Over a decade since LinkedIn's founding, many people are confused about what it is good for. What are you supposed to do with your profile? To what extent is it similar to a resume? To Facebook? And, as importantly, are there ways that you can use LinkedIn that risk offending people inadvertently because you violate a widespread rule of etiquette?

Almost everyone believed there was a LinkedIn etiquette, and part of why someone might attend a workshop on using LinkedIn was to learn what this etiquette might be. But what exactly this etiquette is isn't always clear, leaving me to wonder how a new medium acquires widely acknowledged social rules. At the same time, LinkedIn is purposefully designed to reflect the new metaphor of self-as-business; that is, your LinkedIn profile is meant to be a marketing document through which you can not only showcase your skills, your experiences, and your alliances—all part of the self-as-business bundle—but also continually reveal how you are enhancing yourself. LinkedIn effectiveness and LinkedIn etiquette—these were the concerns that dominated the workshops I attended and the conversations I had with job seekers about LinkedIn, concerns that at the same time address the complicated question of how to operate as a self that is also a business.

Engaging with the Newness of a New Medium

When a new medium is introduced, a widespread etiquette doesn't spring up to accompany it out of the blue. It takes time and work. It is often an open question: Who is responsible for deciding what the etiquette should be? Is it the company? Schools? Government organizations? The users? And which users?

LinkedIn doesn't come with a manual, although there are many online and magazine articles that offer guidelines.[1] And LinkedIn will often make etiquette suggestions through its online help articles, or by sending emails that encourage users to congratulate members of their networks—for their work anniversaries, for their new job, for their birthday. Organizations funded by federal and state governments often provide workshops on how to use LinkedIn. And users will sometimes consult with each other about what to do with their profile.

Yet not all online sites get the same kind of attention from job seekers that LinkedIn does, or become the focus of the same kind of anxiety. Facebook doesn't come with a manual either, and even though I asked about Facebook as well in my interviews, no one asked me, "How do you *really* use Facebook?" LinkedIn, the *professional* social media site, seems to have sparked more concerns about effectiveness and etiquette than Facebook, Twitter, or other sites. When people were worried about Facebook, they mainly discussed their fears that their Facebook presence might be used as a mark against them by those hiring. Job seeking inspires people to delete Facebook photographs or posts, but not much else. In short, different media inspire different types of concerns, even media that are created at almost the same time. And different concerns about media will lead to different kinds of attempts to standardize users' practices.

Every new medium reorganizes the ways that people communicate, changing the participant structure in some way. For example, while emails are often understood in relation to letters, an email's participant structure is different than a letter's in ways that matter. An email indicates who the author of the message is by using an email address that is linked to a password that supposedly isn't shared or hacked. A letter indicates the author through the signature (often handwritten). Obviously, email passwords can be guessed and signatures can be forged. The techniques people use with a particular medium to guarantee that the supposed author is in fact the author can be undercut. An email also circulates differently than a letter does—forwarded instead of mailed or faxed. An email can be sent to others in different ways than a letter can through the functions of reply-all, cc, and bcc. This difference can be significant, as anyone who has had a communication mishap because of reply-all or bcc knows a bit too well. The medium's structure will influence who can be the author or audience for a statement, how many people can be the author or audience, and who is likely to be considered the author or person addressed.

Because each new medium changes a participant structure, a new medium often prompts users to wrestle with the question of how to ensure that everyone uses the medium properly and agrees on what misuse might mean. One of the reasons that this is such a pressing issue is that communication is the result of a complex history of compromises that people have made while trying to share their experiences of the world with others.

Let's take a fundamental example: language. As one linguistic theorist, Benjamin Whorf, points out, language is a set of agreed-on strategies for trying to describe the world.[2] Those strategies are then handed down to future generations. Linguistic expressions have always been compromises. They were attempts to capture a complex reality in words that other people could comprehend. In the process of creating these compromises, grammar and word definitions started to carve up the world. These ways of carving up the world presuppose that the sentences spoken are describing reality. Every communication offers a description of reality that Whorf argues is always affected by the compromises previous language users have had to make. What kinds of compromises about describing the world does language force people to make? Whorf offers two broad examples. First, language always shapes how people talk about the ways that one event follows another, that is, language contains a theory of time. And different languages can contain different theories. Second, language determines what counts as a stable object by defining some parts of reality as things that can be referred to by nouns as opposed to the parts of reality that can be described as a process by using verbs. Every language contains its own theory of the world. But as language is used, this embedded theory is revised as people try to extend what can be said to discuss new contexts and new experiences.

Each medium too is a set of compromises crystallized into a form of communication that travels across contexts. Media, like language, both enable and challenge users as they try to communicate. Unlike most languages (consciously constructed ones like Esperanto or Klingon being the exceptions), media such as LinkedIn have designers who created certain interfaces as they wrestled both with the limitations of computer code and with interweaving many different perspectives about what the new medium should do and look like. So when LinkedIn was released, it was a composite of many different types of compromises and agreements made by people in the company. As people began to use LinkedIn, they had to become familiar with the social assumptions that were built into the design, and many of these social assumptions came from the self-as-business metaphor. The experience of people learning how to use a new medium can be similar to that of adults learning a new language: they often stumble while dealing with the new worldview embedded in the language they are trying to figure out how to speak.[3]

Just like language speakers, users can do unpredictable things with the communicative resources a medium offers. There are limits; these unpredictable communicative acts are being evaluated by their audience, and so they shouldn't be so unpredictable and off the wall that they don't count as communication. With media, you might think that this is something that designers can address. But designers never can predict every way in which a communicative technology might be used. Someone or some group often has to recommend (and sometimes enforce) the rules—perhaps the company's public relations people, or educators, government bureaucrats, or the communities of users, or some combination of all these types.

For example, when the telephone was first invented, people were faced with an immediate dilemma. Telephones provided a new participant structure, which included having to signal verbally who was participating in the conversation and when the conversation was beginning or ending. This led to the practical question: how should the person answering the phone indicate that they are available to start a phone conversation? Edison thought that someone picking up the phone should say "hello," but Graham Bell thought they should say "ahoy." Edison's vision dominated in the United States, although for a while this was touch and go. "Hello" was considered vulgar, and linguist Naomi Baron points out that "as recently as the 1940s, social arbiter Millicent Fenwick deemed the word acceptable only under limited circumstances."[4] In the end, Edison's company was more successful at persuading users to use "hello," partially by including instructions in the front pages of phone books.[5]

Not everyone or every group makes the same decision about how to use a new medium. For example, Mr. Burns, Homer Simpson's boss in the television show *The Simpsons*, still sometimes says "ahoy" when answering the phone. But to communicate, people often have to agree about the general parameters for communicating.

And in establishing these parameters, people are often also deciding on what is signaled by communicating in a particular way, both what is signaled by communicating in the widely accepted way and what is signaled by communicating in an idiosyncratic fashion. Mr. Burns is not only letting other people know that he is old or old-fashioned by saying "ahoy"; he is also indicating something about his class background, and, let's face it, ethnicity. What Asian American would answer the phone using "ahoy"—unless he or she was a Simpsons fan? The cartoonists knew that "ahoy" helped signal that Mr. Burns was a white person, probably from old money, continuing Graham Bell's project. All these signals about identity get wrapped up in the one word someone uses to answer the phone because of complex histories emerging from thousands upon thousands of people answering the phone and telling other people how to answer the phone. This is but one example from one medium. Imagine how many choices that we may now take for granted have been standardized in similar ways.

Histories of Standardizing Media Practices

When a new medium is introduced, how do you get everyone to agree on the basic ways to communicate that should accompany the medium? How do you get everyone to agree to say "hello" when answering a phone call, or at least get everyone to agree that people should say "hello" when answering the phone so that "yo" as the opening response becomes a signal of a certain kind of informality instead of unintelligible gibberish? In the early twentieth century, there were large-scale top-down efforts to teach users what to do: companies, schools, and government offices all tried to teach both social and physical ways to use recently introduced technologies, such as stereographs and telephones. Not anymore. Nowadays it isn't clear that companies believe that everyone *should* share the same expectations around new media the companies provide. How media etiquette becomes widespread depends on the historical period.

When telephones were introduced, companies were deeply concerned about how a telephone conversation's participant structure was different from a face-to-face conversation. The companies decided they had to teach people how to use a party line (with a party line, several households were sharing a single telephone line, which allowed anyone on the party line to join or overhear telephone conversations taking place). Claude Fischer writes: "A common concern of Bell companies, independents and rural mutual lines alike was teaching party-line etiquette. They repeatedly cautioned subscribers not to eavesdrop, both for reasons of privacy and to reduce the drain on the electrical current caused by so many open connections. . . . The companies also tried to teach customers to avoid occupying the line with long conversations. They printed notices, had operators intervene, and sent warning letters to particularly talkative customers."[6] Here the company was involved in instructing users on how to deal with a new participant structure in a number of ways. Companies expected telephone operators to monitor party lines to prevent talkative people (often understood to be women) from dominating this shared medium. Telephone operators (almost all of whom were women) had an assigned role of monitoring as a company representative, so the telephone line allowed not only new, primarily silent participants into conversations but also a new type of participant, the operator, to engage in these conversations and assist in standardizing practices. In the early twentieth century, standardizing practices around telephones and other inventions was not left to the individual user—these newly introduced technologies were accompanied by large-scale educational projects ensuring that everyone was using these technologies in the same manner.

We currently live surrounded by media that have different histories of standardization. Telephones are still with us in a variety of forms. We still use email, and schools have historically taught students how to write a formal email, inspired by earlier lesson plans providing guidelines for writing a formal letter. But this is not true for all the media we use regularly. When new technologies are introduced these days, they aren't often accompanied by etiquette guidelines. Users are often supposed to figure out how best to use a new technology on their own, without much guidance from the company introducing the technology. Contemporary tech companies tend to believe that this encourages users to be more committed to or more involved in using their products.

The contemporary equivalent of telephone operators still accompany new media, but often only as invisible actors making decisions according to a logic of standards that, while they exist, are purposefully kept secret. For example, in early February 2012, the Gawker website posted a leaked copy of Facebook's operation manual for content managers. Facebook had outsourced a task—censoring posted content—to staffing companies like oDesk, who hired freelance workers in Morocco and the Philippines to look at thousands of photographs for a dollar a day. These workers were not told that Facebook was indirectly contracting for their work, but it wasn't hard for them to figure this out. A Moroccan employee leaked the manual he had been given by oDesk, a manual that had been produced by Facebook and that was supposed to help him determine whether a particular photograph was acceptable. The leaked manual was substantially different and more specific than what Facebook publicly claimed were its guidelines at the time. The public version available on Facebook said: "As a trusted community of friends, family, coworkers, and classmates, Facebook is largely self-regulated. People who use Facebook can and do report content that they find questionable or offensive. . . . We have a strict 'no nudity or pornography' policy. Any content that is inappropriately sexual will be removed. Before posting questionable content, be mindful of the consequences for you and your environment." What the operation manual told content managers was forbidden, by contrast, were images of "any OBVIOUS sexual activity, even if naked parts are hidden from view by hands, clothes or other objects," and the manual provided eleven more entries about what might count as a violation of standards.[7] If you only read what Facebook publicly announced about its standards, you would not know that maps of Kurdistan were unwelcome, as were images of earwax, whether real or cartoon, while images of real *and* cartoon snot were acceptable. In short, as a corporation, Facebook felt obligated to hire freelance content managers to protect its image, but unlike when telephones were invented, today many companies are not in the business of openly instructing users about how best to use their technology.

There can be well-known exceptions to this: for example, Facebook wants online profiles to represent offline people,[8] but this is an exception that supports my larger argument that companies are now treating everyone as though they are a business—in this case, a bundle of information that the company can access through a contract. Facebook provides services (such as the Facebook website), and in exchange, the user provides his or her data. For Facebook, much of the value of providing the site to users derives from the data that users offer in return. When Facebook insists in its statement of rights and responsibilities that users "provide their real names and information" and do not "provide any false personal information on Facebook, or create an account for anyone other than yourself without permission," the company is doing more than simply encouraging a media belief in which people are supposed to have only one coherent identity online and offline. When Facebook and other social media corporations try to regulate people's media practices nowadays, it is often in the interest of gleaning information that can most smoothly be sold to others. In other words, Facebook is requiring people to provide information that can be most effectively data mined and turned into profit. What, after all, does an advertising company want with detailed profiles of the kinds of movies a thousand Frodo and Bilbo Bagginses like to watch? In 2012, of the forty-nine posted rights and responsibilities for users, thirty-six of these addressed intellectual property or commodified information in some form or another. So when companies try to standardize user practices nowadays, it is to protect the business and keep it profitable, whether this requires anonymous, invisible independent contractors screening content to protect a company's image or rules linking online and offline identities to ensure that posted information is accurate enough to have value for marketers.

While a new medium may always challenge its makers and users to create some kind of shared understanding of what its use signals, how people tackle this challenge is historically specific. Every new medium may invite users to question how they are extending their already established media ideologies to this new way of configuring the participant structure of a conversation. That is, people may have decided the proper way to compose a letter, and ways to signal whether the letter is more or less formal, meant for a larger audience, or designed for only one reader, and were encouraged to believe this by their teachers in school. When email was introduced, it forced people to rethink all these strategies and decide which strategies used in a letter should transfer to an email and which shouldn't. And so too with more recently invented media: to what extent is a Facebook status update similar to a tweet, and thus subject to the same interpretative expectations?

Because companies, government offices, and schools are no longer as openly involved in instructing people on how to use any new technology, people nowadays often figure this out by talking to their friends, family, and coworkers. For most social tasks, this isn't that much of a problem. If an acquaintance never listens to your voicemails, you may slowly figure this out, and even stop leaving voicemails without deciding that this person is being rude. But there are some tasks that are so highly charged that people pay a lot of attention to what the etiquette might be when new media are involved. Because Americans are surrounded by many technologies in which companies, government offices, and schools have worked for years to establish a widespread etiquette, it isn't that big of a stretch for people to assume that this etiquette does in fact exist, regardless of how new a medium might be, or what the actual efforts to standardize its use might be. Yet historical changes in how etiquette around new media becomes widespread sometimes means that there aren't well-known or uniform rules for how to use some media. This can create an uncomfortable ambiguity for users when these new technologies are part of looking for a job.

LinkedIn as a company provides some guidance to its users about what the company would like to be the widespread etiquette for using this medium and does so more than other social media sites. Yet LinkedIn designers do not necessarily know about all the attempts at standardizing its use. That is, while there are a number of attempts to standardize how people use LinkedIn, these are not necessarily coordinated efforts. I realized this when I met with members of LinkedIn's user experience group. I mentioned to them that some of my research involved attending LinkedIn workshops in the area that taught job seekers how to craft their LinkedIn profile. They were surprised to hear that these workshops existed, and in many cases, only five or ten miles from where they worked at company headquarters. Meanwhile, I was surprised that they were surprised—every community organization for job seekers that I know about offers workshops about creating a LinkedIn profile, and many have been doing so since late 2008, soon after the Wall Street crash.

The LinkedIn employees immediately began wondering if there was anything they could do to help clarify how LinkedIn should be used—perhaps by tweaking the interface or publishing more explanatory articles on LinkedIn, but no one suggested coordinating with the people leading these workshops. I left the meeting wondering what would happen. Would LinkedIn employees remain committed to this belief that online design decisions and articles could be enough to shape how people decide to use LinkedIn for complicated social interactions? As Graham Bell found out, it is one thing to come up with the rules for using media, and it is another thing entirely to convince everyone that your rules are the ones that should be followed.

LinkedIn as a company tends to focus its efforts at standardizing users' practices to the way it designs its interface and to making advice available online or in offline publications. Like Facebook and some other sites, it encourages users to fill out their profile as much as possible. Yet LinkedIn takes this a step further by showing a circle gradually filled to indicate the ways in which someone can have a profile operating at "full strength." This graphic can encourage people to fill out as much information as they can on their profiles, an effective strategy for creating a certain degree of standardized use, or is it? I saw this strategy in action one day when I was observing Ruth, a career counselor, offer a free consultation to Gina, which she did for anyone who wanted to improve their LinkedIn profile. Gina's first concern was how she should fill out the portion of the LinkedIn profile in which she described her job title. She was unemployed, and so she had no job title. But, as she pointed out, to have the LinkedIn Profile Strength meter confirm that her profile was at "All-Star" status (the LinkedIn terminology), she needed to fill out this section. Ruth thought that Gina should not take the LinkedIn Profile Strength meter all that seriously. However, Ruth suggested she could put down something in that space if she found it comforting, and recommended that she put down "actively seeking opportunities in [a specific professional field—in Gina's case, biotech product management]."

These are precisely the conversations by which many social media standards get established. The interface urges the user to behave in one way, yet the actual user's circumstances don't allow him or her to easily or smoothly accede to this suggestion. For example, LinkedIn has an implied user who would fill out the profile completely, but perhaps the actual user doesn't have a job title, as in Gina's case (this is but one of the ways in which the LinkedIn interface tends to presume that its users have jobs). Or, in another example, Don did not want to indicate what year he graduated from college in case that marked him as too old to participate effectively in this particular job market. So the actual users check with friends, coworkers, or a local career counselor to see how to finesse the gap between the implied user and the actual user's context and practical concerns. And as they do this, they come up with solutions that feel appropriate to that group. Over time, this can turn into an agreed-on etiquette for people who are all in a certain profession, or in a certain region of the country, or of a certain class background. Even when LinkedIn's interface strongly recommends a standardized practice, there is no guarantee that users will comply unless it makes sense in terms of how their group communicates about jobs and employment already.

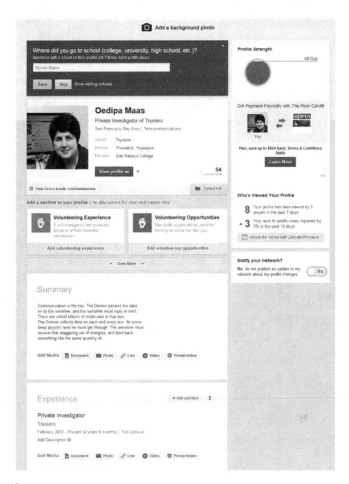

Figure 3.

Unemployed and Connected

Job searching is a social task that, in the United States, is so highly charged that people often worry that there is a clearly defined right and wrong way whenever they communicate with a potential employer. So what are people's common concerns about the right ways to use LinkedIn?

When people first start using LinkedIn, they often are uncertain about how to decide whom to connect to, since a LinkedIn connection is an element of the LinkedIn participant structure that is specific to LinkedIn. Connecting with someone gives them a certain kind of access to your profile, although the exact degree of access depends on how you set your privacy settings. People commonly talk about only connecting with people they know—although what counts as knowing someone can be quite varied. Some will insist that they have to know the person for a certain amount of time. In a focus group I conducted with job seekers, Thomas explained that now that he is looking for a job, he meets quite a number of other job seekers in various job search workshops and meetings. If they request to connect with him on LinkedIn, he will wait to find out whether he sees and talks to them again. He explained: "There's a bunch of people now that I'll look back and they've sent a request after I met them at one event, it's two months later and I have no clue who this person is. I can see their face and I still don't remember who this person is." For Thomas, LinkedIn connections signal a potential obligation. Someone might ask to be introduced to one of his LinkedIn connections, and he will have to decide if he feels comfortable making the introduction. He wants LinkedIn connections to mean more than simply exchanging business cards. Others, however, see adding someone to their LinkedIn contacts exactly as though it is an exchange of business cards.

People who use this more expansive strategy for connecting are quite clear that their LinkedIn connections are potentially not as useful as those of people who are more discriminating. They have many more contacts, but the people they can contact are less likely to be willing to respond when they ask for a favor. This was a problem that job seekers mentioned frequently about LinkedIn connections—a connection symbolizes potential, but precisely what kind of potential? What will happen if you do in fact ask a favor of someone whom you primarily know through a LinkedIn connection? Will the person do what you ask? If you are willing to connect with anyone who asks, you may have more connections than people who are less inclusive, but they might not be all that valuable as connections. Knowing someone only or mainly through LinkedIn contacts was seen as a weak tie indeed. It was a pretty common belief that the more contact you had with someone that wasn't mediated through LinkedIn— perhaps in person, by phone, or by email—the more likely that person would be to perform a favor, even if you asked for the favor through LinkedIn.

When I talked to college students about their strategies for connecting to someone on LinkedIn, they often were trying to understand whether to use for their decisions the same criteria they had developed for choosing to connect with people on Facebook. One student, Kate, was debating whether to connect with someone she didn't much care for. If the request had come through Facebook, she definitely would have found some way to avoid connecting. Kate explained that she would refuse to connect with anyone by Facebook with whom she might be ambivalent about sharing personal information. But this request was sent through LinkedIn. She was uncertain how to value a LinkedIn connection, largely because it was on a professional networking site. This person might turn out to be useful in the future—it is hard to tell when you are twenty-one who in your circle of casual acquaintances will turn out to be a connection you want to maintain in the future for professional reasons, regardless of what you think about them at that moment. And, let's face it, some people who are unpleasant at twenty-one may learn to be decent by the time they are twenty-eight.

Leslie, another student, was uncertain whether she should connect with all the people who requested connections with her at the company where she was interning. She had never met many of these people. They were just requesting a LinkedIn connection because they worked at the same place. If these were Facebook requests, Leslie would have turned them down immediately. She wants to at least meet someone in person before connecting on Facebook. But a LinkedIn connection? How should she screen? She was clearly working at a company in which people were connecting without any actual interaction, a local workplace practice that I did not often come across. Most people whom I spoke with, of all ages, would only connect with people whom they had had some interaction with, even if it was only by email or in a LinkedIn group.

While most people I spoke to tended to be more guarded about their Facebook connections than their LinkedIn connections, I did talk to one woman in her midforties who is far more careful with her LinkedIn connections than her Facebook ones. She, like the others I spoke to, views Facebook as personal. But for her, personal means that she is ready to connect with anyone from her high school, or anyone whom she has some kind of personal connection to, however vaguely defined. She rarely goes on Facebook, so she doesn't value the connections highly. She sees Facebook primarily as a site for broadcast communication, for communicating with as many people as possible at once. She much prefers one-to-one communication for personal interactions. Whenever possible, she talks to people on the phone, and so a Facebook connection is relatively meaningless to her, a polite acknowledgement that there was some kind of personal relationship at some point in the past, even if it was decades old.

But LinkedIn is different. She sees it as a professional site that reflects her reputation. If she connects with someone on LinkedIn, it is a sign that she is willing to recommend that person to someone else if asked, and so she carefully vets her connections. When she first started using LinkedIn, she was only connecting to people she knew through a second job she had as a massage therapist. She felt that as a massage therapist, her recommendations to others in comparable but adjacent professions—acupuncturist or nutritionist, for example—contributed to her own reputation in this line of work. She saw connecting with someone as tacitly recommending them to others and didn't want to connect with people whom she couldn't vouch for. Facebook connections, from her perspective, carry no such endorsement. Yet it is important to keep in mind that the entire time she has been using LinkedIn, she has had a primary and stable job. She has never faced the pressures to change her approach to LinkedIn connections, pressures that job seekers face from the moment they start actively looking for a job.

Searching for a job tends to encourage an approach to LinkedIn connections that is focused on increasing connections that can be used instrumentally, and to inspire many more standardized and widespread strategies for when and how people connect on LinkedIn, much more so than other social media designed these days. Yet just because there is more standardization doesn't mean it happens smoothly, or that LinkedIn as a company gets to decide how it will be done. Actually connecting, in fact, is a moment in which career counselors strongly encourage people to refuse LinkedIn's own efforts to standardize this interaction. LinkedIn provides its own language whenever you want to request a LinkedIn connection with someone. In 2014, the company provided the sentences "Since you are a person I trust, I wanted to invite you to join my network on LinkedIn" or "I'd like to add you to my professional network on LinkedIn." By using *I*, each implies that it is a sentence the author of the profile has in fact typed, instead of what the sentence is—a template that LinkedIn provides. Indeed, in the first example, LinkedIn's word choice announces that users actively trust the person they want to link to. With this phrase, LinkedIn is also encouraging users to think about LinkedIn connections as an indication of trust, not potential usefulness.

Counselors believe that to use one of these prepackaged sentences signals that you did not take the time yourself to write a personalized sentence. People will interpret the invite in various ways when they see it couched in language they immediately recognize as supplied by LinkedIn, regardless of the sentence's claims otherwise. Some people genuinely don't care whether someone has personalized a LinkedIn invitation to connect. Other people view this gesture as crass. If they are feeling generous with their time, but don't recognize the person inviting them to connect, they might write back asking for the clarification that they think *should* have accompanied the invitation. With this in mind, career counselors usually encourage people to alter the wording, to signal that the person is attempting to craft a more personal connection by replacing LinkedIn language with language that calls attention to the context in which people first met or the reasons for requesting this connection. This is an instance of clashing attempts at standardization, with LinkedIn and career counselors guiding people in different directions.

But do these clashing suggestions have any effect on what people actually do? True, some counselors who recommend this didn't follow their own advice when requesting to connect with me on LinkedIn. But others put considerable effort into making sure they supplied their own words in their LinkedIn invites. Yet LinkedIn's interface doesn't always give you the option to change the wording of your invitation. In 2014, it was impossible to personalize a request to connect if you were using the LinkedIn app on your cell phone. In those instances, LinkedIn always supplied the language used to connect; users had no alternative. Some people were aware enough of this as an issue that they refused to connect with someone using the LinkedIn app and waited until they got back to their home or work computer.

How Public Should You Be?

In most workshops on how to use LinkedIn, some new adopter would ask: how public must my profile be? This question is about participant structure—who is the audience of a given profile? LinkedIn, like other social media, allows you to control this to a certain degree. Many in the workshops were already familiar with Facebook and policed their Facebook privacy settings. Yet people teaching LinkedIn classes would strongly recommend that job seekers be as publicly accessible on LinkedIn as possible. Job seekers should want to be found. Although when counselors declare that of course job seekers want to be found, found by whom remains a bit ambiguous. In practice, LinkedIn tends to be a site where job seekers are found by recruiters, but not all jobs are ones that employers hire recruiters to fill. Sometimes, being public on LinkedIn is useful only to signal that you want a job.

People in the LinkedIn classes often expressed significant reservations when told that they ought to be so public with their profiles. Some were worried that people they had conflict with in the past—disgruntled coworkers or angry exes—would find them on LinkedIn. I even heard about a woman who refused to go on LinkedIn because she had adopted a child and was now concerned that the birth mother would change her mind and insist on getting her child back. In these instances, people felt pressured to be as public on LinkedIn as possible in order to find a job, but they also were trying to address other social problems in which being too public on the internet might lead to unwelcome social consequences in their daily lives.

In discussing the issue of publicness on the web, those leading classes on social media and LinkedIn often made the argument that this was an issue of control—that there is a tremendous amount of information about you available on the web by now, and it is best to be in control of that information. Posting information yourself was supposed to ensure that you were in control of what people could find out about you on the web, as I discussed in the context of personal branding.[9] And while I heard this claim that you could have control often enough in workshops, I have to admit that this logic baffled me. It does seem to be based on a fairly typical way that Americans understand how the meaning of a statement is determined—that what the author wants the statement to mean will dominate how the statement is received. That is, if you are in an argument and someone shows you a text, tweet, or email you wrote, and you claim "that is not what I meant to say," then your intention when writing should determine what that text, tweet, or email meant. But anyone who has gotten into an argument over what a text message actually meant, or whether a sentence was sarcastic, knows that in practice, the author of an utterance tends not to have much control over how it gets interpreted. And how often on the internet does someone get to clarify? Job seekers don't often have opportunities to talk about their intentions with the recruiters or people in HR who are looking at and interpreting their LinkedIn profile or history of tweets. I was not the only one not entirely convinced by the energized claims that actively choosing to be public was giving you control. While those teaching the classes were often quite enthusiastic about the control you can have over your self-image, those in the classes were not so easily convinced. Given what they said in response, they didn't seem to experience circulating information on the web as a liberating moment in which they have complete license to shape their image however they want.[10]

People are now expected to be public job seekers in a way that they had never been until online job boards such as CareerBuilder and Monster emerged in the mid-1990s. Yes, people from the eighteenth century onward might have placed newspaper ads announcing that they were looking for work.[11] But these job ads contained very little information about who the job seeker was. Online job boards changed this, turning resumes from being a document with a limited and predictable circulation into a document that circulates unpredictably and very publicly. Until online job boards, people would only send resumes to companies. Job seekers understood that this meant a certain loss of control over how resumes circulated. Companies might keep the resumes on file, and the candidate couldn't know who precisely at the company would look at the resume. HR would probably be involved, as would a hiring manager, but the job applicant might not know the names of anyone who had looked at his or her resume until the job interview. Even then, the resume could have circulated more widely within the company than the applicant realized. Yet there was still a relatively limited audience for any resume. Online job boards changed this. Resumes became public documents accessible to anyone who stumbled upon the document online. The process of looking for a job encouraged people to reevaluate how they understood a resume might circulate. People became resigned to making their work histories widely known. LinkedIn continued this practice, and in the process, it created a database of resumes that recruiters have found very useful as they search for likely candidates for job requisitions they hope to fill.

While Monster and other job boards might have helped people get accustomed to having public resumes, resumes on job boards and LinkedIn's profiles are different enough from traditional resumes that new social dilemmas arise from having a public resume or a public LinkedIn profile when looking for a job. One of the issues that any job seeker who isn't a recent graduate struggles with is how public you should be on LinkedIn about the fact that you are looking for a job. People with a job aren't always comfortable letting their boss or coworkers know that they are looking for a job, and they tend not to mention it on their public profiles. While posting your resume on a job board indicates that you are looking for a job, having a LinkedIn profile does not automatically mean that you are looking. It is an ambiguous signal. Although one of the implications of the self-as-business metaphor is that you are always potentially on the verge of leaving your job, there are still professional consequences for indicating this publicly.

People who are unemployed also often are ambivalent about whether they want everyone to know they are looking for a job. Does it help their chances of getting a job to have everyone know that they could start immediately? Or is the prejudice against unemployed people so great that it will hurt their chances of getting a job? There are two main places on your LinkedIn profile where this is an issue: in your headline (the first four or five words that appear directly under your name on your profile) and where you identify your current place of employment. Some people won't announce publicly on LinkedIn that they are no longer at their former place of employment. Some people will create consulting companies, literally claiming to be businesses themselves, so that they have a company name to put in that slot. Others are certain that availability will make them much more attractive to recruiters, and signal this with phrases like "looking for new opportunities."

Some people view the public nature of LinkedIn as an opportunity to tantalize recruiters with just enough information to persuade the recruiter to contact them, but not enough information to get the recruiter to quickly dismiss them. People occasionally discussed writing a LinkedIn profile with just the right balance of information so that recruiters weren't sure whether they would be a good match for the job. Mario, who had gotten a job recently, explained: "One of the challenges is making sure you don't get eliminated based on what someone reads. You should only have enough information there that they reach out to you. You want to open up a dialogue rather than making someone think that they've already read everything about you." Mario thought that recruiters were just as likely to use too much online information to screen out applicants as they were to become interested. Others would tell me that they believed too much information about all the jobs they had would be confusing and would lead the recruiter to believe that they did not have the necessary skills when they in fact did. To address this, Mario and others tried to write profiles that enticed but did not inform. This was a careful guessing game in which you had to predict how much would intrigue an unknown recruiter to think you might be a possibility, without giving away too much information. Here the LinkedIn profile is being used to anticipate one particular type of audience—recruiters—and, unlike what LinkedIn designers believe, users intend to create a profile through the careful art of concealment and omission in the hopes that withholding information will inspire recruiters to request offline revelation.[12] Because LinkedIn profiles are so public, these users have developed a new set of strategies for this genre different from those they use for offline resumes.

While many people I knew carefully omitted some of their work history on LinkedIn, I did hear a recruiter in a workshop enthusiastically recommend listing every job. She thought everyone should try to place as much historical information on LinkedIn as possible and be more circumspect with the resume. This recruiter's logic was that since LinkedIn is a public profile for creating as big a professional network as possible, you want to let everyone from your past know how to find you. If you worked someplace in 1993 and you don't mention it on your LinkedIn profile, your coworkers from that period have less of a chance of finding you through LinkedIn. Not everyone has the same take on what a LinkedIn profile does or how others will interpret information on a profile.

Endorsements and Recommendations

Endorsements are a feature of LinkedIn that manages to condense many of the social quandaries that this new participant structure offers users. Users are asked by LinkedIn to endorse other users' skills by clicking on a button to affirm that, say, Robert is a good public speaker. Endorsements were an adaptation of a trending database that LinkedIn offered so that users could figure out what words describing skills were most popular in the jobs they were interested in that month. They could track how the mention of certain words increased or decreased.[13] In 2012, LinkedIn decided to transform this "skill" database into a lightweight recommendation system— one-click endorsements. They provide an unqualified positive affirmation of people's skills. There is no way to discuss how good you are at something, just that you can do it. I might be able to create a PowerPoint presentation, but not a compelling or memorable one. On LinkedIn, people can testify that I can in fact produce a Power-Point presentation when called on to do so, but the endorsements don't allow them to nuance this claim. I could be a PowerPoint drudge or grandmaster—LinkedIn endorsements only indicate the presence of a skill.

Anyone can endorse a profile for a skill. Tony explained to me with great energy how frustrated he was by his well-meaning relatives who kept endorsing him for skills on LinkedIn. His aunt Mary was terribly worried about him, and she kept endorsing him on LinkedIn even though she had no idea what he did as a product engineer or what the skills actually referred to. Tony wanted endorsements to be legitimate. He believed that the person clicking an affirmation that someone possesses a certain skill should have some previous knowledge of the ways that person has demonstrated this skill. Tony kept having to monitor his LinkedIn profile, and delete endorsements, because his well-meaning relative's endorsements couldn't possibly signal substantive knowledge. For many people, endorsements seemed a bit suspect, a signal, yes, but a fuzzy one and not necessarily a legitimate one.

Some people would use endorsements simply to spread goodwill in the world. Luke explained to me that he would endorse anyone whom LinkedIn suggested he endorse, operating by the philosophy that he would accept people at face value unless he was actively forced to confront their lack of ability. Of course, this is LinkedIn's assertion of face value, not something the people themselves actively claim. Luke explained to me the logic he used to decide whether to endorse people after I asked him why he had decided to endorse me. We had met briefly at a workshop and connected on LinkedIn, and he immediately endorsed me for university teaching. I was puzzled; he had never seen me teach. This was even before we met for an interview, and I was just beginning to learn that people had different views about endorsements. Luke explained his take: "My basic philosophy is, you put your hand out to shake hands. I'm gonna shake your hand. I'm certainly not gonna bite it off. To me, endorsements are like that. I'll give out endorsements freely to people. You tell me that you're an experienced professor, in anthropology, I believe you. OK, you're an honest good person. You tell me you're ethical, I believe you. And that's what endorsements are to me." In his attempt to spread goodwill, Luke would get on LinkedIn every day and spend ten or fifteen minutes endorsing the people whom LinkedIn recommended. Luke was such a prolific endorser in job-seeking circles that other people would occasionally mention him to me as an example of an endorser gone wild. Will, a fellow job seeker, mentioned: "I ran into him last Saturday and he goes, 'Excuse me, have we met?' I'm like, 'You endorsed me five times.'" In this instance, neither Will nor Luke thought endorsements were consequential, but Luke saw the endorsement as fundamentally about reminding people of your goodwill toward them. And Will was amused that Luke's use of endorsements didn't mean Luke would be able to remember anyone he endorsed when he met them in person.

At the time of my fieldwork, LinkedIn's interface encouraged people to act like Luke and to endorse as a game-ified form of connection. When you first logged on to LinkedIn, the interface would present you with the opportunity to endorse four people you are connected to through LinkedIn, and it also suggested the skills that you should endorse them for. Once you clicked an agreement to endorse all four, you would be presented with another four, and so on until you decided to stop endorsing people. LinkedIn would prompt you to endorse people for specific skills; you tended not to decide on your own (for example, "Oh, I should really endorse this person for public speaking").[14] For many of the people I spoke to, this aspect made endorsements suspect.

To endorse someone was to give information that seemed to have little value and only contributed to LinkedIn's ability to keep people engaged on the site for a bit longer. How and when people were invited by the site to become an endorser shaped how they evaluated what being an endorser meant and how seriously they interpreted other people's acts of endorsement. In short, a LinkedIn endorsement is an illustrative example of a new technology offering a new participant structure, which results in the accompanying social dilemmas people face with any change in participant structures, namely, of how to evaluate this new version of authorship, content, and audience. And in job searches, many people wish there were clear and widespread ways of interpreting something like an endorsement, a standardization which Whorf might point out takes work on many people's parts to achieve.

In addition to endorsements, recommendations turn out to be a relatively charged aspect of LinkedIn. In 2006, LinkedIn changed its interface so that your LinkedIn connections could post a couple of sentences attesting to how good you were at doing your job. Yet just because it is possible to have recommendations on your profile doesn't mean that they are easy to accumulate. When job seekers and career counselors are talking about LinkedIn recommendations, they are often focusing on how to most effectively and politely request a recommendation. People often discussed writing their own recommendations, or sketching what they would like to see in the recommendation, and then asking their former coworkers or bosses to fill out the sketch. So job seekers see getting recommendations as a complicated enough social request that they consult with others about the best way to go about getting them, and they worry about controlling the wording. But their strategies also mean that the authorship of a recommendation may be a bit more up for grabs than LinkedIn's interface implies. While LinkedIn's interface clearly states who the author of the recommendation is, this doesn't always mean that this is the actual author. Time and time again, I heard about situations in which authorship in practice was not so clear—about instances in which the person receiving the recommendation had at minimum initiated the post through a request, and often had written it. Sometimes people were faced with very busy possible recommenders. Sometimes they were concerned about how fluent their potential recommender was in English. In sum, the actual participant structure shaping a LinkedIn recommendation is not necessarily the implied participant structure.

Even in instances when people write LinkedIn recommendations for others out of the blue, there is often an instrumental logic behind this. And this can lead to clashes in interpretation. Job seekers will occasionally begin writing recommendations for their former coworkers in the hopes that this will inspire them to reciprocate. Indeed, this form of recommendation implied a social reciprocity for many people. One woman, Maia, who had a job as a software engineer, explained to me that she would never ask someone to write a LinkedIn recommendation for her if she wasn't positive that she would be willing to write one for them. Yet I also talked to Susan, a recruiter who actively screened out candidates who had recommendations that reflected this form of exchange: "I look at how many recommendations you have. And more important is: are they wimpy or are they reciprocal? If I have a suspicion, you know, if all of them are kind of wishy-washy, and I can see that they are all reciprocal, you know that with this person, something is wrong. You can have one or two, we can both really respect each other, but when all the referrals are reciprocal, that's a huge red flag." Susan wanted some way to determine whether a recommendation was genuine, and she saw mutual exchange of recommendations as a signal that the recommendation itself might be suspect. While for Maia this exchange was a sign of mutual respect, for Susan it was a sign of an inappropriate and suspicious quid pro quo exchange. Here Maia and Susan simply interpreted the participant structure they saw revealed through LinkedIn recommendations differently. This tension over different ways of interpreting participant structure was a relatively invisible concern socially—Maia and Susan were unlikely to be in a position to encounter each other and experience this clash. But almost every other dilemma I heard about LinkedIn revolved around people's openly voiced concerns over perceived ways that LinkedIn might be affecting the participant structure of communication, concerns that were mainly raised in workshops on LinkedIn or in my interviews.

LinkedIn and Offline Social Ties

There are not only ambiguities about how to use the participant structures LinkedIn offers. There are also challenges in figuring out how and when LinkedIn reflects the ways people organize themselves offline. Part of what LinkedIn offers is a way to connect with people solely on the basis that the two of you were part of the same organization at some point or another—that you worked for the same company or graduated from the same college or were part of the same sorority. But what does this commonality actually indicate? One recent college graduate I interviewed told me how difficult she found it to connect with others who had been a member of her sorority on another campus. I was surprised, as sororities are supposed to be ideal organizations for creating the networking ties that so many job seekers want to establish. She explained that each sorority has its own character on a college campus, and that the reputation of a sorority might be specific to that campus: "With different sororities from campus to campus, they could be a top house there and just a really weird house there. It kind of varies, and so to say that I would have a connection with this woman because she is in my sorority does not mean anything." Here it is a question of what kind of tacit information is conveyed when she connects with a sorority member: what stereotype might she unwittingly be engaging with? This might seem to be a problem limited to only fraternities and sororities, but company offices in different parts of the country have their own workplace-specific dynamics, as do international branches. Being part of PayPal in the United States might signal something very different than being part of PayPal in Spain. For that matter, a recruiter explained to me that the years you were hired at a particular company might also convey additional information to others. She said that one company had a reputation for making terrible hiring decisions between 2002 and 2005. If you had been hired at the company beforehand, other companies would be happy to poach you. But having been hired during the bad years at the company could be a mark against you, even though working at that company in general would not be. Institutional reputations, from sororities to corporations, can be complex and require considerable background knowledge. While LinkedIn offers a way to search for people who at some point or another have all participated in similar institutions, it turns out that how similar these institutions are in practice might be crucial (and sometimes worryingly unknown) for people as they decide to connect and exchange with each other.

The specificity of a workplace can cause problems for LinkedIn users in other ways too. For example, the playful connections you are supposed to enjoy in Silicon Valley workplaces don't travel well onto your LinkedIn profile. I came across one example early on in my fieldwork, in which I interviewed someone at a start-up company where every worker had a goofy and nondescript job title such as flight leader. Two days after I heard Sean describe these job titles, the *New York Times* published an article discussing how members of start-up companies had begun to harm their chances at job transitions by using job titles that were too quirky on their LinkedIn profiles. Apparently, if recruiters happened to come across their profiles, the job titles were so opaque that recruiters would simply go to the next potential candidate.

A few weeks later I was speaking with Sean again. He had decided that it was time to quit, and he was describing how he was preparing to look for a new job. So I asked him how he was changing his LinkedIn profile in anticipation of this. He immediately mentioned changing his job title, so I brought up the *New York Times* article. Sean responded: "If you go to my LinkedIn right now, I don't put any of that gobbledy gook. Because it's confusing, it screws up the algorithm for LinkedIn based on how it curates and helps people search and find you. . . . Most people asked me what a Flight Leader is, which was the first red flag. Second red flag, because you get an email from LinkedIn if you sign up for it with weekly jobs that are related to you, I was getting stuff from Boeing."

Sean initially used the job title on his LinkedIn profile to address his coworkers, to signal an inside joke, and he was not trying to present himself in a way that made sense to a more generically professional audience. As he began to reorient himself away from his coworkers and toward an audience that might help him find his next job, he edited his LinkedIn profile to make it more legible for that potential audience. He started to exert control over how he represented himself that signaled his individual future plans, not his connections to others at work. In short, he changed the audience being addressed by his profile in part by his choice of job title, and in doing so, he changed the participant structure called up by his LinkedIn profile. In this instance, Sean was responding to the signals he was getting from both LinkedIn and casual conversations about how these audiences might interpret his statements.

Second-Order Information

Part of what people were concerned about is not what was actually said on LinkedIn but what was implied by using LinkedIn in a particular way. When a new medium is introduced, one element that is often up for grabs is how its form affects the message. This was particularly true in job searches. Most of what I have been describing throughout this chapter is anxiety about the second-order information that a message can convey. Second-order information is information that the medium, or form of the message, conveys on top of the actual words or images used in the message, information that changes as the medium's use changes. For example, a cell phone's area code can provide information about where someone was living when they first got that cell phone, but this is not the same information as a landline's area code, which indicates the general geographical location where the phone call is in fact being made. The second-order information you learn from a cell phone's area code is part of the caller's geographical history. You can believe that this information indicates quite a bit about someone's behavior or style of interaction, or you can believe that it indicates very little. You might think that anyone who has lived in New York City talks quickly and drives aggressively. Or you could believe that the only thing that you learn from discovering that someone has a cell phone number with a New York City area code is that they lived in New York City at some point in their life.

Time and time again, I was struck by how often job seekers worried that employers would reject them based on what I considered second-order information—that is, that employers would reject them based not on the actual content of the application but on the form of the message. One example that was occasionally discussed in the workshops I attended was that having an email address with an old domain name such as "aol" or "hotmail" could be grounds for rejection. These domain names supposedly marked an applicant as out of date. Career counselors would recommend having a Gmail account to indicate that you are adopting the more current technology. True, while I heard this advice often enough, I never met anyone who admitted to me that they screened job applicants based on their email account.[15] But this anxiety was indicative of how everyone in the process understood that job applications were not only evaluated by what was said; they were also evaluated by the second-order information accompanying the various genres submitted for evaluation. Job evaluators will take aspects of how people are using a medium and what medium they are using as indications of what kind of worker they will be—although I want to stress that they won't pay attention to the same things or interpret the same second-order information consistently.

I came across plenty of other instances in which hiring managers or HR managers admitted that they screened people for how they used media to communicate. One woman in HR told me that she would check people's Facebook profiles regularly to see how they presented themselves. She did not care whether they drank alcohol, but she did think twice if she saw too many photographs of the person drinking. For her, this was an important glimpse into the kinds of social judgment this person exhibited. She did not want to hire someone who was too careless about their self-presentation, and she saw Facebook photographs of a person drinking or wearing risqué clothes as a warning sign that he or she might behave indiscreetly in other contexts as well. She also believed that the decisions people made in setting up their Facebook privacy settings provided valuable insights into someone's discretion, and she would notice this as well. I also heard people complain about the fact that their coworkers would take *anything* about an applicant's Facebook profile as a reason to reject a job candidate. Since not everyone agreed on what posting something on Facebook indicated, this could become an obstacle as people tried to come to a hiring decision together.

This is part of the frustration of job searching. People are using second-order information constantly to evaluate job candidates, but it is difficult to predict which aspect will matter, especially given the wide range of second-order information that every applicant is submitting with every job application. A tremendous amount of job-seeking advice is aimed precisely at suggesting ways to standardize the signals you might be sending through your choices. Following this advice homogenizes your applications, making it more difficult for an evaluator to select your application for rejection.

But this standardization is a double-edged sword. It also makes it more difficult for your application to seem distinctive, and thus to be selected. This is why, despite all the advice they receive, job applicants will try to do something unusual, like using infographics on a resume. They are trying to find the right balance between being so clearly competent at the genres of a job application that they won't be rejected and distinctive enough that they will be noticed.

Hiring managers are aware that they are often making very quick decisions based on this second-order information. One explained to me how he dealt with the large number of resumes he received for one job posting:

> So we got three hundred resumes. I can have someone filter through them, but I go through them myself. . . . I get a pile of them and I look for a reason to reject them. And as soon as I find it, I can go on to the next one. And it's like a relief. . . . And then if I can't *immediately* find a reason to reject them, then suddenly a switch turns in my head and I'm looking for a reason that I want them because I want the whole miserable process to be over.

He understood that he was not reading resumes in a generous light, or trying to see the potential in every candidate. Instead, he was primarily reading to say no — and these noes often depended on quick judgments about how people represented themselves based on relatively little information. But in this case, generosity was only a few seconds away.

Hiring managers recognize that all the information they have is already prepackaged to put the applicant in the best possible light. And some candidates will look more plausible than others, even with only a quick glance at the resume. In short, on the hiring side, anyone evaluating applicants is trying to narrow down, on average, fifty to three hundred job applications to the two or three most promising candidates. The work of going through a pile of applications is not trying to figure out the best in each candidate. It is weeding out the rejects as quickly as possible.

When a new technology that people might use for hiring is introduced, job seekers immediately start worrying about how to use it so that they don't get rejected. When they worry, they are acknowledging that hiring is a social activity in which people can easily get rejected because of employers' unpredictable interpretations of second-order information. One career counselor told me that a client discovered he hadn't gotten a job as a sales manager because he had only 100 LinkedIn connections. The hiring manager believed that anyone in sales should have at least 250 connections.[16] The hiring manager had decided that the number of LinkedIn connections someone has indicates their skill as a salesperson, although readers can probably come up with several reasons why this wouldn't be an accurate indicator, including the possibility that the salesperson found other ways of staying in touch with potential customers to be far more effective. In short, new technologies introduce many new opportunities to circulate second-order information, and in hiring situations, this means introducing many new reasons to accept or reject a possible candidate. If you use these technologies in nonstandard ways, you risk being rejected. But how do you determine what is the standard way for media that are new? Or what happens when new technologies have too many uses, say, shifting at unpredictable moments from being used for social reasons to being used for hiring purposes, as many worry happens with Facebook?

Because hiring managers and recruiters all too often are looking through resumes trying to winnow down their list of possible candidates, job seekers will become anxious about the right way to use a new genre or medium for presenting themselves as hirable. Every new technology shifts the ways in which second-order information is circulated. In situations in which every detail has the potential to be scrutinized, any small shift in the second-order information circulated could matter far too much. For example, LinkedIn profiles suggest you list contact information. Do you list your email address? Molly Wendell, in her advice book on job searching, strongly recommends that no one list their email address, because this signals that they are too eager to be reached by anyone who happens upon their profile.[17] Yet other career counselors will just as strongly advise people to list their email address so that they are easy to contact — why put obstacles in the paths of interested recruiters? Job seekers who don't want to announce too clearly on LinkedIn that they are looking for a job might make their email public to encourage recruiters to think that they could be tempted by the right offer. Companies tend to hide their employees' emails online so that they can't be poached by other companies. To list or not to list? The question is not about the actual email address but the second-order information signaled by listing. Providing your email indicates a willingness to be contacted. The disagreement lies in what that willingness implies. If everyone automatically put their email addresses on their LinkedIn profiles, not having your email address would be a different statement, one about your willingness to conform. Without a standardized etiquette, the question becomes how people will interpret your willingness to be accessible by email.

For other new media, such as Facebook, this problem is more acute. To a certain degree, because LinkedIn is understood to be a professional medium, users borrow more closely from other previously standardized forms, such as the resume. As companies, both Facebook and Twitter provide far less overt instruction than LinkedIn does. Of course, they are also being used to accomplish a much broader range of social tasks. Job seekers are addressing a number of different audiences on Facebook, Twitter, and other similar social media—they are interacting with their many different kinds of friends and family at the same time that they are trying to anticipate how a potential employer might interpret their Facebook or Twitter communiqués. Not surprisingly, nowadays many college students begin to concentrate on cleaning up their Facebook profiles when they are seniors and starting to think about getting jobs. But they don't always know what to delete and what to keep. Gail, a senior at Indiana University, explained to me that she was removing all the photographs on her Facebook profile that might potentially be viewed as a sign of poor judgment, including her Halloween photos in which she dressed as a Teletubby. A Teletubby is by no means a risqué costume, but Gail was worried that she might seem too quirky. She wasn't sure what exactly she should be concerned about, so she ended up worrying about every sign on her Facebook that she might be marching to the beat of a different drummer. I interviewed a significant number of people who had this generalized anxiety, in part because the only guidelines they found for using particular media came from talking to friends and family, or coming across news stories about people who got fired for a Facebook post. Companies and governments aren't involved in creating standardized practices around the introduction of new media as they have been in the past, and so figuring out how to use these new media is even more confusing than previously.

As people develop techniques for using and interpreting Linked-In's participant structure, they are often borrowing from the other genres that LinkedIn profiles resemble. People might describe LinkedIn profiles by pointing out how they differ from resumes. Career counselors would explain that LinkedIn profiles, for example, are supposed to be spaces where people narrate the history of their work experiences with the first-person singular; unlike resumes, LinkedIn profiles are supposed to be filled with sentences that begin with *I*. What a LinkedIn profile is supposed to look like is determined in part by what it resembles but is not—similar to a resume, but not; similar to Facebook, but not. In describing what LinkedIn profile photographs should look like, people were constantly tacitly or openly comparing these photographs to Facebook profile photographs. One man explained to me his view of what an acceptable LinkedIn photo was:

> The biggest mistake people have is they have a nonprofessional photo in their LinkedIn Profile. I've got a folder that I call bad LinkedIn photos that I just acquire as I run across these people. I've got pictures of a dude wrestling an alligator. I've got pictures of a PWC consultant barefoot skiing. I've got a picture of a guy with his two kids climbing over his head like he's a jungle gym and they are climbing over his head. My point to those people is that LinkedIn is where you do business. If you were going to make a big sales presentation to Wells Fargo about their back-end banking employment system and you were going to walk into the boardroom

with the board of Wells Fargo sitting there, would you let your three-year-old kid climb up to your head with a bunch of chocolate pudding dripping down your shirt? Because that's the picture you've got on LinkedIn.

These comparisons also allow people to develop increasingly sophisticated understandings of what kind of second-order information is being circulated by the different media. When you learn how to interpret a resume or a Facebook profile, you bring these techniques to a LinkedIn profile. In the process, you often begin to notice whether LinkedIn's interface reveals the same kind of information as the other media you use, and begin to distinguish how you use different media.

At the same time, when you interpret a LinkedIn profile, you are doing this in the context of the other ways a person is representing him- or herself as a desirable worker. And the contrast can define how a profile is interpreted. If you are comparing a LinkedIn profile to a resume, you will notice different things than if you are comparing that profile to a Facebook profile. How you interpret a genre is influenced by the rest of the genre repertoire available, and thus implicitly, those other histories of standardization.

More people have asked me what LinkedIn is good for than have ever asked me what Facebook is good for, or what email is good for. Part of the problem with answering that question is that it separates the social from the technological. What LinkedIn is good for is what people performing a social task with you find it good for, and how they, and you, evaluate the communication that happens through LinkedIn. If you are a recruiter, you might find LinkedIn good for looking for people who use the keywords that also appear in the job requisition that you are trying to fill. If you are a job applicant trying to find someone you know in a company you are applying for, and if people in that company answer LinkedIn requests (and that is a big "if"), then LinkedIn can be a good way to contact someone. If you are a job applicant researching a company, trying to understand who is in a company and what they do, and you look at people's profiles, you are often reading profiles written for another audience. These are profiles that are often written for recruiters or future employers who might want to consider hiring the person when he or she wants to leave the company. So the profiles are written as generally as possible, with an eye to the next step in someone's career. In those moments, you will have to interpret the profile imaginatively to figure out the information that you care about. In short, what LinkedIn is good for depends on how different communities use it.

Figuring out how to use LinkedIn when you are searching for a job condenses many of the issues faced by anyone using new media for a highly charged social task. You have to learn how the medium you are using relates to the other media used for similar tasks, such as resumes or business cards. You also have to learn how the medium is unlike other media that might appear similar, such as Facebook or Twitter. In learning how this particular medium is distinctive, you are figuring out the ways in which that medium configures participant structures—the kinds of roles it enables, the ways authorship and audience are shaped by the medium's structure, and how people interpret information gleaned through the medium.

Yet there are aspects about figuring out how to use LinkedIn "properly" that revolve around LinkedIn specifically, and especially the ways in which LinkedIn is meant to be a certain kind of solution to a gap in job seekers' genre repertoires. LinkedIn, after all, provides a new genre to this repertoire when no new genre has been added for decades. And it does so in part because the older forms of representing your self as employable all are infused with the logic of the self-as-property metaphor. LinkedIn allows you to display your self as a business, a marketing document in which you can present your self as a bundle of unweighted skills (endorsements), unweighted relationships (connections), and experiences. This is a bundle that LinkedIn would like users to believe they must constantly manage and enhance—especially since social media companies measure their success to a certain degree by how often they are used. Part of the second-order information that the LinkedIn interface reveals is how effective users are at adopting this model of the self, of presenting themselves as an ever-expanding businesslike bundle of skills, experiences, and connections.[18]

158   At the same time, users' struggles with LinkedIn also reveal some problems that emerge when you try to implement this model of self-as-business. The LinkedIn version of the self-as-business consistently errs on the side of being context free. You can see this in the ways endorsements are unequivocal: you either have a skill or you don't, but how you learned the skill, when you demonstrated it, and to what degree you might possess it—all of these things are unknowns to anyone trying to interpret your LinkedIn endorsements section. It is the same with LinkedIn connections. It is unclear to anyone looking at a profile how well the profile owner knows those he or she is connected to. I have described some of the etiquette issues that arise for users when they are dealing with a genre that tends to signal quantity instead of quality, especially when they are looking for a job or a potential employee, a moment when the context matters tremendously. As LinkedIn provides a platform for representing your self as a business, it errs on the side of being more of a marketing document than an evaluation, enabling formulaic ways of expressing your self as a bundle instead of enabling you to provide nuance or context.

Five

Changing the Technological Infrastructure of Hiring

I have been talking a lot about how resumes and interviews have changed to reflect a new understanding that people own themselves as though they are businesses—collections of skills, assets, qualities, experiences, and relationships that continually have to be managed and enhanced. Yet some people feel that these hiring genres haven't changed nearly enough, and that older understandings of employment still shape the way people hire, to everyone's detriment. They want to transform hiring practices even further and are creating new platforms that more effectively reflect the business-to-business model of employment.

There are many ways to think about problems in hiring and in workplaces. I have been arguing that the self-as-business metaphor shapes people's social analysis of hiring problems and affects the kind of advice they give and seek. I have been talking largely about the everyday social strategies people come up with to solve hiring dilemmas and exploring some anthropological and sociological concepts that are useful for evaluating all the circulating advice. The people whose stories I tell in this chapter take a very different view. On the whole, they think hiring is a broken process because of *technological* reasons, and they are trying to find technological solutions.

Admittedly, I was researching hiring in Silicon Valley, a place where technology is generally seen as the solution for every problem. But what happens there matters. Last decade's technologies were in fact attempts to solve perceived hiring problems: companies like Dice, Monster, and LinkedIn are providing online tools for managing the hiring process in response to gaps that those companies' founders wanted to address a decade ago. And they have transformed the landscape of hiring across the country. Now new start-ups are trying to fix what they perceive as the sociotechnical problems of today. Discontent with hiring in general, and often LinkedIn in particular, has motivated a new technological scramble. Perhaps these new technological solutions will solve problems of hiring. But I'm skeptical. I think it's far more likely that today's technological solutions themselves present users with social quandaries that they will then have to develop workarounds to address.

I interviewed the founders and early employees of a few of these companies established after 2011 to understand the philosophy behind the products they offer. Many of their hiring solutions are their attempts to build parts of the now dominant self-as-business metaphor into the hiring process more effectively than the founders think resumes or LinkedIn profiles are able to do. The people designing these new platforms talked openly about how their companies are trying to solve problems in the hiring process. I was one of many people hearing their spiels. These founders are constantly in a position of having to explain to potential investors, clients, prospective employees, and journalists what makes their companies distinctive. Many think resumes are a particularly ineffective genre for presenting people's skills, so some focus on the problems of using resumes to select potential applicants. One company thinks that adding infographics to resumes will improve them. Another company wants to create software that will be more effective at sorting through resumes and selecting appropriate ones. Other founders criticize the managerial-class assumptions built into technologies such as LinkedIn and have chosen to design platforms for the working class. Some want to change the social dynamics that seem to undercut the ways hiring should, in their view, be a meritocracy, or based on skills. These founders feel that too much hiring is based on who someone knows or where someone went to school. Behind all these solutions lies the same impulse—these companies want to change the infra-structure and practices underlying hiring so that hiring can more accurately reflect not only a self-as business model but the partic-ular version of the self-as-business model that the founders believe in. Many problems the company founders see with contemporary hiring spring from the ways current technologies and genres presup-pose older understandings of employment or the wrong version of the self-as-business model.

This, to be clear, is my interpretation of the inspiration behind many of these companies. No one talked to me about helping people to act more like a business, or helping them to avoid acting like ear-lier historical models of employees. To the degree that anyone talked about the past, they focused on how inadequate they found previous technological solutions. This is not surprising, as every new technol-ogy invented now to solve hiring problems is partially responding to earlier technological solutions.

For example, applicant-tracking systems loom large in some of these CEOs' imaginations. Applicant-tracking systems, after all, were in part designed to deal with the new influx of applications employ-ers receive now that they can circulate job descriptions online and thus reach a much broader set of job seekers than a local newspaper ad might attract. Some of the company founders I talked to (and most job seekers) feel that applicant-tracking systems are flawed, screening out perfectly reasonable candidates. These founders want to create better screening software—thus solving a problem created by an earlier technological solution.

Some companies are trying to supply an alternative to other avail-able platforms, such as LinkedIn. LinkedIn, after all, has built into its interface an assumption that the user is a white-collar worker, some-one who has a job or hopes to get a job as a white-collar worker in a corporation. LinkedIn allows users to present a certain type of work history—one that focuses on what office workers might want to share about themselves as a bundle of skills, experiences, and relationships. The site isn't designed with tradespeople in mind, as I explain in more detail when I discuss the company WorkHands. I also am not LinkedIn's imagined user. The interface is not currently built for academics, as I remember every time I log on and think about my professional needs. After all, what I and many of my academic colleagues most want to do is circulate our own academic publications, not necessarily circulate other people's well-written and thoughtful blogs and news articles. LinkedIn is not currently the go-to site for scholars to find people's latest academic journal publications; I and my colleagues visit other websites to do this. Similarly, the LinkedIn interface doesn't work well for people working on short-term projects like television commercials, as I address in my description of DoneBy, a company offering a professional media platform designed to solve some media professionals' career-specific networking problems.

All technologies, not just LinkedIn, presuppose certain types of users. And when they are presupposing a certain type of user, they are also presupposing the social relationships that the user has.[1] In LinkedIn's case, the platform is presupposing the kind of work the user does and the best way to explain succinctly to others how that work can be translated into a series of accomplishments. The further you are from this imagined LinkedIn user, the harder it is for you to use LinkedIn for your own aims.

Many of these companies have different approaches to how best to fix what they see as the problems in the hiring market, problems created by a mismatch between older ways of circulating hiring information and this new model of self-as-business. Some companies think that the tools currently available are too generic, are designed for a standardized office worker, and have built-in social assumptions that are inaccurate. In two of the cases that I discuss, DoneBy and WorkHands, according to the founders, the current hiring technologies do not adequately deal with the specific ways in which some jobs are temporary, such as work in the trades or in movie and television production. Other companies recognize that not all aspects of the self-as-business are as easily measured. You might be able to provide metrics for your skills or relationships (1,250 LinkedIn connections or 800 Facebook friends), but not your personality. Companies such as Knack offer ways to quantify your qualities, reconfiguring a decades-long history of psychometric scholarship for this new model of work. Meanwhile, companies like Entelo, Gild, TalentBin, and others believe that networks and certain experiences — such as college degrees or working previously for well-regarded companies — are all too often valued far more than people's skills in the hiring process. They develop software that they believe allows employers to privilege skills above the other aspects that compose the self-as-business. Finally, the founder of TalentCircles believes that companies and job seekers should know about each other well before a job comes open, so that they can be more knowledgeable when they finally do enter into a business-to-business agreement.

Ingredient Branding

John Gibbons, one of the founders of DoneBy, used to work for IMDb (the Internet Movie Database), which allows users to figure out all the films a director has made, or all the filmed productions in which a given actor appears. Film and television crew members would occasionally contact him at IMDb, asking him if IMDb might be tweaked to list all their experiences making commercials or short videos for YouTube. After enough of these requests, Gibbons started to wonder if he could design a platform that would be more useful for camera and boom operators, key grips, and all the crew working behind the scenes on multiple projects. Crewmembers are hired for a specific project as individuals, not as a team. Sally, the key grip, will easily lose touch with Harry, the boom operator, until they happen to work on another project together. And it might be difficult for a director to track down the cameraperson for a commercial or perhaps another kind of project that would not appear in IMDb. Gibbons wanted to create a platform that would allow people to be easily found for their many different media projects, to show examples of all their work on their profile, and to stay in touch with the crewmembers they had worked with in the past. Using this website, even if Sally forgot Harry's name, she could easily find out who the boom operator was that had worked with her on that Diet Pepsi commercial in 2005.

DoneBy is providing a solution to a problem that can crop up when you believe personal branding is important. To explain what the problem is, I have to return for a moment to the idea of branding, and in particular the issue of ingredient branding. As Robert Moore points out, objects face particular problems being recognized and having a distinctive brand if consumers only come across them as ingredients in other products they consume or use. For example, NutraSweet is part of Diet Pepsi, and its delivery is dependent on that beverage, making it hard for those branding NutraSweet to make consumers realize that they are not only drinking Diet Pepsi, but they are also drinking NutraSweet.[2] But you can find packets of NutraSweet that aren't already mixed in with another product. Intel, a company that makes semiconductor chips and microprocessors for Apple, Dell, Lenovo, and other computer companies, has a more serious ingredient branding problem. The company struggles to distinguish its components from the computers that contain its products. Certain jobs ensure that an employee will face the same dilemma. You might recognize a Woody Allen movie, but how many know that it is also edited by Alisa Lepselter? And since Alisa Lepselter has been editing every Woody Allen movie for over fifteen years, she has a better chance than most of being recognized.

DoneBy is offering a new way for someone to become known online for the work they do in a certain professional context where the way in which the work is organized determines both how people establish a reputation and why they might need a reputation in the first place. It allows people to make their labor visible in a new way, collecting all their projects in one place, projects which on their own don't publicly acknowledge the crew labor that produced the media product. After all, commercials don't have film credits at the end. This in turn allows people to overcome a common branding problem that objects have long faced, and that people are only now beginning to wrestle with in this specific way as they begin to think of themselves as also having a brand.

Social Media for Blue-Collar Workers

The company WorkHands was founded out of a similar impulse: to create a platform that responds to the employment-specific demands of jobs that no social media site serves well at the moment, in this case, workers in the trades. One of the founders explained that he first got the idea for WorkHands when his brother sent him a text. His brother had just spent six hours working on an underground electric vault and wanted to show someone his accomplishment before it was buried. His first thought was "That's really cool!" And then it occurred to him that this photo could be much more useful to his brother than simply as an image to share with family and friends. He realized that many blue-collar workers don't have an easy way to show potential employers what projects they have done in the past. Yet now that so many people have cell phones with cameras, people are carrying cameras with them on most work sites. It is relatively simple for them to take photographs that can function as a portfolio of their work, but they need a site to collect the information in a way appropriate to their work communities. He recognized that sites such as LinkedIn have a white-collar employee as the implied user, and a different interface would be necessary for blue-collar workers if they were to use a work-specific social media site. He wanted users to immediately feel comfortable when they go on the WorkHands website. "I was talking to a glazier," he explained, "the people who do glass on buildings, and he was saying: 'I went and signed up for LinkedIn, got through the whole sign up process, poked around a little bit, and my gut reaction was—this isn't for me. All the pictures are people in suits. It's all like tell me about your various educations.' It just didn't fit with the world he lived in."

At the same time, WorkHands' philosophy is still very much in dialogue with this new model of worker. Both WorkHands and the next company I discuss, Knack, turn to the self-as-business model and explore whether some aspects of the bundle—skills, assets, experiences, qualities, or relationships—can be represented more effectively, perhaps visually or through metrics. The founders of WorkHands realized that the current forms for documenting what an electrician or carpenter has done in the past are woefully inadequate. But in finding a streamlined way to represent a person's skills and previous experience online for workers who build or fix things, WorkHands also enables these workers to portray themselves as a bundle of skills and experiences.

In trying to encourage workers to adopt this new model, the WorkHands designers ran into a problem. The workers using their site weren't connecting with other people they had worked with on various projects or at different construction sites. Collecting work alliances through a professional online profile didn't make much sense to them. They didn't stay in touch from project to project, and they weren't exactly sure why they would want to do so, in part because they could rely on unions to help them find work. What was making intuitive sense to people on filming crews didn't seem like a pressing need to people in the trades.

The designers at WorkHands wanted to encourage these connections as another avenue to circulate information about possible jobs, but they were then faced with a problem. What do you call these connections? They aren't friends, so Facebook's terms wouldn't transfer. Calling them connections resonated too much with a white-collar way of speaking that didn't feel comfortable in this context. The founder I was interviewing explained: "So we say, let's just turn to the experts, put it on our Facebook page, and help us settle a bar bet, what do you call the people you work with? And the number one answer was 'asshole.' It was twenty different people put it in there, half of them got six likes and things like that, and so we actually flirted with the idea of calling them like 'these are assholes I know' but decided against that." The WorkHands founders were struggling with yet another consequence of how new media create new participant structures. When the roles are too new—such as an online tie that is anchored in a certain offline practice—it becomes hard to figure out what terms to use when naming the new role. They settled on "contacts."

Both DoneBy and WorkHands are creating websites intended to address perceived problems with how workers in project-based jobs can represent their varied experiences over time and sustain connections with each other. Both companies are imagining a space without unions serving the function of connecting workers. They are both addressing structures of how work is organized in these professions that have existed long before the self-as-business metaphor dominated people's vision of employment. Yet the solutions to the perceived problems are shaped by the metaphor, as they encourage people to imagine themselves as a bundle of connections and experiences that can lead to further work. Before the metaphor, people might have imagined solutions that involved visiting union halls regularly or other ways of creating solidarity. With the metaphor, the solutions revolve around turning to online platforms to manage and enhance people's professional connections and record their work experiences.

Measuring Your Qualities

While DoneBy and WorkHands provide a certain segment of workers with the opportunity to present their experiences in a new way, Knack enables people to measure and list their personal qualities by taking cognitive tests in the form of video games. Guy Halfteck, Knack's founder, realized, after a particularly frustrating experience trying to get a job at a certain company, that companies have no good way to determine what a person will actually be like in a job. Resumes, interviews, cover letters—none of these genres are particularly good at revealing someone's personal qualities. Personality tests abound, including the ubiquitous Myers-Briggs test, but most of these tests can be gamed. If you know in advance that you want to be an INTJ (introvert, intuitive, thinking, judging), you can answer the Myers-Briggs questions in such a way that you will get the result

Figure 4. Dashi Dash, a video game and cognitive test for determining people's personal qualities. Image courtesy of Knack.

you want. Halfteck wanted a personality test that did not seem like a personality test, one that allowed people to enter into a state of flow and concentrated immersion while they were being tested. He turned to video games, believing that playing a video game could be so captivating and unusual as a cognitive test that someone could easily become absorbed in the game's puzzles. Players could easily move beyond that double consciousness of both thinking about the personality questions and knowing that how they answer will determine their future working life.

When I was doing this research, Knack had only three video games/cognitive tests that you could play to discover your qualities, or, as the company calls them, your knacks. I was most intrigued by one of these games, Wasabi Waiter, now called Dashi Dash, because it is the only game in which one animates a human, and the only game set in a work context. When you play Dashi Dash, you are a waiter in a sushi restaurant who has to serve a varying number of customers as quickly and efficiently as you can.

You don't have to deal with any coworkers, no sushi chef, no other waiters—it is just you and a handful of quickly discontented cus-tomers. Why a sushi restaurant? At first, the Knack team designed a bar and a bartender, but alcohol seemed too problematic a substance if this was going to be used globally. The team needed a setting that would not force them to completely redesign all that they had cre-ated up until that point, so they settled on a sushi restaurant. The customers could sit at a counter, lined up in the same way people sit at a bar. And sushi was becoming so widespread that the team figured it would be a recognizable food and setting in many different countries.

I have to admit some of the things that I found fascinating about choosing a sushi counter as a setting did not seem to occur to Guy Halfteck or his team when I raised the subject with them. I was struck that they were choosing a setting in which someone was play-ing a person relatively low in the labor hierarchy—being a waiter in a sushi restaurant is not many eight-year-olds' dream job, or many twenty-eight-year-olds' dream job for that matter. It is also a job that requires a certain kind of emotional skill. To cajole customers takes a different kind of emotional labor than managing a global team of computer engineers, or convincing a jury your client is innocent. And the emotional work involved in dealing successfully with cus-tomers is different in turn than the social knowledge it takes to get along well with your coworkers or your boss. All my instincts about the game were to focus on how the context might involve very spe-cific techniques for managing relationships, social interactions that those playing the game may never have had an opportunity to prac-tice, or may have had too much experience practicing. For me, the social narrative was all-important; for the game designers, not so much. They were far more interested in how the ways someone had their avatar act in the game could be translated into metrics for per-sonal qualities.

Halfteck himself, like other start-up founders I spoke to, sees his company as promoting meritocracy in hiring. He believes that where people graduated from should not determine what jobs they get. He sees accurately measured personal qualities as a better and more honest guide to whether someone would be good at their job. As he outlined what Knack could become, he began to talk about how a child growing up in a favela in Brazil might discover that he or she had exactly the right combination of knacks that the most successful people at Google or Shell had, and this would encourage that child to apply to those companies. He sees uncovering people's knacks as a way to uncover the kinds of jobs they might be good at doing, jobs that they might not otherwise have dreamed of trying to do.

I asked Halfteck if Knack could also reveal who was likely to behave badly—if Knack might allow employers to implement the "no-asshole rule." I was not the first person to ask him this. He explained that companies were too different internally for Knack to be able to do this accurately. The personal qualities that made someone successful at one company would be different from what made an employee successful at another company: "When we capture the knack signature for successful people at that culture, at that company, it captures both what makes you successful as an engineer and what makes you also successful culturally in terms of matching and thriving in that environment. So when we look at successful engineers at Microsoft, the fact that they are successful and have been around at Microsoft several years is because, not only because they are able to do their job well, but they are also able to work with others well." Yet the qualities that make you seem like a great team member in one company culture can make you seem like an asshole in another company culture—it depends on the company. Halfteck points here to how qualities might in fact interact with context, offering the perspective that different workplaces enable different configurations of qualities to thrive or fail. In this perspective, one company's asshole is another company's inspirational leader.

In providing an easy way to measure and name personal qualities, Knack addresses one obvious deficiency in the way that the current genre repertoire for hiring inadequately represents the self-as-business model. Knack after all finds a way to quantify in very legible terms what has previously been difficult to measure and difficult to represent through the available forms—personal traits.

Yet Halfteck's vision for how people will use Knack implements the self-as-business model in another way as well. Halfteck wants Knack to be used to help match companies and applicants. He would like to see every company get its most successful employees to play the video games and discover their knacks, creating a Knack profile of what kind of person is likely to be successful at that company. Once a company has its own company-specific profile, job seekers will be able to determine if that company is in fact a place where they can thrive. Halfteck wants Knack to reveal not only a person's qualities but also a company's favored personality, so to speak. This would allow job seekers to assess their potential fit at a company, rather than those on the hiring side being the only people determining fit. Job seekers are imagined as canny choosers, not anxious or desperate to get a job, any job. Knack is yet another attempt to make more likely the promise of equality that currently is a dormant potential in the business-to-business metaphor underlying contemporary hiring.

Enhancing Recruiters' Rolodexes

Some of the start-up companies I researched were consciously offering recruiters a very different database of potential employees than LinkedIn, trying to provide potential candidates who had been vetted by other online communities. LinkedIn, after all, is a site that is designed to allow people to promote themselves, so it errs on the side of highlighting the networking and branding aspects of the self-as-business. Unlike other forms of evaluating work and expertise available on the internet, LinkedIn only allows endorsements that are all or nothing, as I mentioned earlier. If I want to endorse a friend for his or her teaching abilities, I can't indicate that my friend is marvelous at lecturing, but mediocre at leading discussions, and downright disgraceful at returning students' exams and papers on time. I either have to endorse him or her for university teaching or refrain. Recommendations listed on people's profiles are carefully curated by the profile holder. This issue came up occasionally as people would tell me about instances in which they found out a coworker was misrepresenting his or her work experience on LinkedIn. The former coworker might claim to have had a different work title than he or she actually had, or to have been the leader on a project when this wasn't the case. There is no way to comment on the accuracy of another's LinkedIn profile. It is a social networking site that, in the process of connecting people with each other, does not allow for the monitoring and disciplining that communities invariably do. And thus LinkedIn profiles are generally understood to require a particularly skeptical type of reading.

At the time, I found four companies in particular that attempt to provide recruiters with a different set of insights into potential job candidates than LinkedIn does. Entelo, Gild, TalentBin, and Swoop-Talent all provide recruiters with software that creates potential hires' profiles by scouring the web for all publicly available information on a specific person. Of these companies, only SwoopTalent explicitly advertises that its software is useful for catching candidates in lies and misrepresentations in their applications. The company explains on its blog that it can check whether people really do have the credentials they claim, or determine what the actual relationship is between candidates and their recommenders: "Checking each candidate's references can be a time-consuming and sometimes frustrating task. You may call a listed reference only to discover the person only worked briefly with the candidate and does not remember much about him or her. Some candidates even recruit friends or family members to pose as professional references, despite your job application instructions not to do so. Running both names through social sourcing will often uncover the truth about references."[3] While only one of these companies might be openly marketing to companies who distrust job seekers' representations of themselves, all these companies are compiling online communities' public evaluations of each other's skills to find potential employees.

Most of these companies are focused on the tech industry and find potential candidates by scraping information from websites like Stack Overflow or GitHub and other open source communities. Stack Overflow is a site where computer programmers post questions for others to answer, and these answers are rated by the rest of the Stack Overflow community. Every user has a quantified reputation. The more others like your answers, the better your overall, numerically determined reputation is. The more votes "up" your post is given, the higher your reputation's number. The more people vote "down" your post, the lower your reputation. Thus Stack Overflow as a community is evaluating its members' answers in easily quantified ways that can be interpreted as evaluating someone's skill.

In practice, someone's skill and someone's Stack Overflow reputation are not one and the same. This, after all, is a community, and all sorts of social factors contribute to how people vote on answers. Sometimes people will want to support their friends, or others who have voted positively on their answers in the past. Sometimes people are responding to a style of presentation or history of interactions built up over time. And in the case of GitHub and other open source platforms, many people have been pointing out that women can find these communities inhospitable. Just as in any other social community, reputation is a social product as much as it is a general consensus about the presumably objective value of answers or information.

These companies in particular are not trying to fix the perceived inadequacies of the genres job seekers use to present themselves as employable. They are trying to avoid the genre repertoire altogether. They hope to use information provided on the web as a supposedly transparent lens to find potentially desirable candidates. Some of these companies, such as Gild, will claim that these techniques encourage more diversity and a more genuine meritocracy. After all, they find candidates based on their web interactions, not based on where they went to school or how old they are. Some companies, such as Entelo, will claim that their algorithmic analysis of web practices gives a better insight into when someone is considering leaving their job—they claim to have found the right way to read the tea leaves of online activity to see when passive candidates are willing to actively consider moving on, indicating to recruiters that now may be a good time to approach them. In short, these companies are transforming the internet itself into one large job application, claiming they can see through its lens sharply to read someone's skills unhampered by the messiness of someone's life history, and to read intentionality unhampered by the messiness of someone's offline contexts.

Changing Participant Structures

When I first began to do research on changes in hiring, I started to look at different US cities' government websites for the unemployed. New York City's website, like others, provides links to many other websites, some government-sponsored, some private. As I was following various links on the New York City website, I came across a casual mention that companies need to be concerned about their recruitment brand. This stopped me in my tracks. A recruitment brand? Companies need to be concerned that their image as a company extends not only to how they treat their customers but also to how they treat their potential employees? I wondered how this might play out in practice. Was the idea of a recruitment brand powerful enough to encourage companies to treat job applicants with consideration and respect? I have to admit that this was the only mention of recruitment brands that I found in many months of doing fieldwork. I knew that if I suggested to job seekers that perhaps, just perhaps, one of the changes in hiring that had been happening over the past thirty years was that companies felt an increasing sense of corporate responsibility to job applicants, they would look at me as if I was delusional. For the most part, consideration and respect for job seekers was thin on the ground, and having a thick skin for being treated shabbily is a necessity for people actively looking for work these days.

Yet the idea of a recruitment brand made me wonder: if everyone involved in hiring now genuinely believes that they are entering into a business-to-business contract, would that make companies change their perceptions of job seekers and how best to treat them? On the surface, it seems like this change in metaphor would have the potential to do so. After all, the metaphor suggests that both parties are equal members, and that businesses need to treat their employees and potential employees as well as they treat the other businesses they interact with that provide them with much-needed goods and services. In addition, if people are now supposed to move rapidly from one job to another, a job applicant may not be someone you want to hire for your current job opening, but he or she may be a great candidate for a future job. Thinking about a company's hiring practices in terms of a recruitment brand suggests that a company should pay attention to how its hiring practices are perceived, and try to monitor how applicants become convinced from the moment they consider applying that the company is a good one with whom to enter into an alliance.

What if one of the problems in contemporary hiring is that companies want to encourage employees to see themselves as businesses when it is in the company's benefit to do so, but not when it is in the job seekers' benefit? Perhaps if this metaphor were applied more consistently to everyone involved, applicants would be told when a job has been filled, or accurately informed about how long it will take for a hiring decision to be made? Even these relatively small gestures would go a long way to helping job seekers feel that they are being treated equitably, and as though their time and effort in applying is recognized as valuable.

It wasn't until I started talking to Marylene Delbourg-Delphis about her start-up company, TalentCircles, that I heard anyone talking about recruitment brands again. She was very clear that she was trying to transform the hiring process by providing a platform that changed the conversational roles and structures of interactions involved in hiring (what I have been calling *participant structures*). Early in our interview, she pointed out that hiring hasn't changed for decades. She explained why she thought this was: "The reality is that if you are a job seeker, you just basically experience the same problem as you did ten years ago, you go to a career site, you stay informed, and then you hope that somebody will get back to you, and in practice nobody does. It's a small minority of companies getting back to the customer. In the hiring process, you attract people, you get them to the career site, and then they fall into the unknown, into the abyss—this has not changed." For Delbourg-Delphis, the difference between a LinkedIn profile and a resume is insignificant, so too the difference between an online job application form and a paper one. Applicants still feel as though they are at a serious disadvantage, kept from having all sorts of information that they would find useful about how the hiring process is going. Recruiters still have the same set of tasks that they had before the web: creating a pool of job applicants for a single job, and selecting a short list of likely candidates through conversations with the hiring manager and a series of short one-on-one screening interviews. What Delbourg-Delphis decided to create is a platform for companies to adopt that enables them to create relatively long-term relationships with a pool of potential employees.

What does this new participant structure for hiring look like practically? A company has a social media site in which it posts information about itself. If someone is interested in working for that company eventually, they create a profile on the company's site. They might do this in response to a specific job ad, and they can apply for any job at that company through the TalentCircles platform as well. But people can also create a profile on the site because they are interested in the company in general, not because they definitely want a job there that month. All someone is doing by signing up on a company's site is indicating their potential interest in working for that company.

These sites are company-specific, and they are not all interlinked. A person can sign up for as many company sites as they want, and the different companies will never know what other companies the person is interested in flirting with. Delbourg-Delphis herself will never know. She considers all the content that populates the platforms— the users' profiles, the companies' posts—to be the companies' intellectual property, not hers. Unlike LinkedIn or Facebook, TalentCircles only sells the platform and has no overarching access to all the data input into the platform.

TalentCircles has another feature which can be used to change the participant structure of hiring as well. While hiring managers and recruiters can conduct individual video interviews with potential job applicants, they can also conduct group interviews with job applicants. Readers may be familiar with group interviews in which one job applicant talks to a number of people already employed at the company. TalentCircles inverts that participant structure: it allows an interviewer to have a wide-ranging group discussion with a handful of job applicants. As Delbourg-Delphis explained, a friend of hers was trying to sort through one hundred resumes from undergraduates, and they all seemed the same: "The resume of one student is the same as a resume of another student." She suggested that her friend invite the applicants, eight at a time, to have a conversation in the chat booth. Her friend could watch everyone talk together as they tried to solve a problem for thirty minutes, and pick the leader in each conversation for a follow-up interview. Instead of interviewing one hundred people for a month, her friend could screen all the applicants in six hours.

This significantly changes the participant structure of a screening interview, making the recruiter primarily an observer of group dynamics, and having the interview itself shift from being a question-and-answer session to a free-ranging conversation among relative strangers in the same structural position, in this example, recent graduates looking for a job.

These video interviews and group discussions are recorded, so that hiring managers can also watch, should they choose to do so. As Delbourg-Delphis pointed out, currently recruiters take notes on each screening interview and report back to the hiring manager. These interviews aren't recorded (although some companies are beginning to offer this possibility). The recruiter is functioning as an intermediary, translating and re-presenting the job applicant's words to the hiring manager. With this platform, the hiring manager can choose to change the recruiter's multifaceted role in the process, allowing the recruiter to be the interviewer and go-between only, and not also the translator. Of course, this will take up the hiring manager's time.

In providing a company with its own social media platform for potential job applicants, TalentCircles is significantly transforming the roles of the company, the recruiter, and the job seeker in the hiring process. When Delbourg-Delphis explained this to me, she focused on how TalentCircles changes what it means to be a recruiter:

> One of the assumptions in my platform is that recruiters are marketers. I would simply summarize this in saying that recruiting should be part of marketing because the recruiter is basically the person talking to the outside just as a marketing manager is talking to masses. Even a recruiter is a one-on-one marketer. If the recruiter has a bad image, chances for me as a candidate to buy the products of the company are cut like 50 percent. People have to realize that that position is a department of marketing and in fact HR management should be a part of that. The idea that the recruitment department is run by itself is completely asinine. It was not the case in the old days and the idea is absurd. As you fill another position, you are the spokesperson for the brand. Even more than that, the personal brand [here Delbourg-Delphis refers to the recruiter's personal brand] is speaking to individuals or communicating with a group of individuals. So you cannot sell them with canned messages.

Delbourg-Delphis is focusing on how the recruiter becomes largely responsible for the company's recruitment brand, mostly because of how her platform is structured.

And Delbourg-Delphis also could be changing when and how companies have to pay attention to their recruitment brand. You could claim that every company's online job application form is an instantiation of its recruitment brand, and indeed, in the one other mention of a recruitment brand that I found while doing this fieldwork, the online job application portal is precisely what the anonymous author of the website focused on. Yet TalentCircles offers a way to present a recruitment brand differently, through all the ways the company interacts through the TalentCircles site with those who have self-selected themselves by joining as potentially interested job seekers. But who precisely is doing this interacting on the company's behalf? Delbourg-Delphis clearly believes it will be recruiters, and that this process will change their job description substantively, encouraging them to be constantly attempting to maintain company followers' interest in the company through their own selected words. On behalf of the company, they are supposed to create personal relationships with those willing to be in the reserve pool of potential employees.

At the same time, if the platform truly catches on, this might ensure that recruiters would no longer be valued for their rolodex, so to speak. The professional connections with possible candidates that recruiters develop working for one company could no longer travel with them as easily when they start working for another company. Instead, the connections would be more effectively cemented into the platform through which companies make contact with potential applicants. Since, as I mentioned earlier, recruiters are often on temporary contracts with companies, and often transitioning between companies more frequently than those they hire, this platform has the potential to alter how a company assesses the value of a recruiter. It would not surprise me to learn that the TalentCircles platform encourages companies to insist that recruiters animate a generic avatar on the platform. Bob the Recruiter might be animated by any number of different actual recruiters over the course of the company's lifetime, but the potential applicants might believe that they were consistently interacting with one person.[4] This could create a much needed appearance of continuity for a company, while minimizing the potential loss of an individual recruiter's knowledge when his or her contract is up. It also has the potential of demanding that recruiters learn how to excel at maintaining fictional personae through different channels of communication.

Yet job seekers too have their roles transformed through the structure of this platform. At first glance, it might seem like those who join TalentCircles are agreeing to be a reserve supply of labor, waiting patiently for the company to choose to hire them. The site might seem to resemble the street corners and parking lots where temporary construction workers or gardeners line up, hoping an employer in a pickup truck will drive by and ask them to work that day. Yet because this is online, there is a significant difference. Potential employees can join several different companies' sites, and no one will ever know (in part because companies will not share this information for fear of helping their competition). Job seekers have more options when they are no longer limited by having to physically show up to see if a job will come their way. Being in this particular type of reserve pool has the potential to give job seekers more possibilities, which in turn forces employers to devote time and effort to keeping these candidates potentially interested in working for that company. At the same time, being in this particular reserve pool could take more time and effort over a longer period of time than showing up on a corner, thus limiting the number of companies that one job seeker can realistically interact with. In this sense, TalentCircles has the potential to change the type of effort and amount of time companies and job seekers devote prior to the moment a job offer is made and accepted.

TalentCircles changes the participant structure of hiring interactions so that it more readily reflects the idea that hiring is a temporary business-to-business relationship, in which both parties need to be actively persuaded to enter into an alliance. In the process, this potentially changes the roles of all involved. Recruiters would become primarily marketers, actors, and skilled observers. At the same time, job seekers would be more explicitly encouraged to choose five to eight companies for targeted job searches, and applying to a given company would require greater time and effort than tailoring a resume for a specific job description and researching the company once the candidate has an interview. In addition, immediate job openings may no longer be the primary basis for interacting with a company as a potential employer. People would be strongly encouraged to choose the companies they want to work for, and then hope a job they fit becomes available. But all this is speculation. When one changes the participant structures of how a process takes place, people will engage with the participant roles newly made available to them in imaginative and unpredictable ways.

Designing a solution for a hiring process always involves tackling a number of tasks at once. To design a new platform, you have to explore how people evaluate each other and through what forms. You have to think about the genres people use to reveal information about themselves, and see if the genres reveal information that will help those hiring make a decision. You have to think about the ways in which those hiring reveal information throughout the hiring process, helping job seekers decide if they want that job and if they want to join that company. It means asking if these forms of evaluation can be done in ways that are more in keeping with the ostensible interests of the job seeker, or the company, or the hiring manager, or the small group of coworkers who will work with the newcomer—all of whom have their own often clashing interests in the hiring process.

When these start-ups design technologies to change hiring, they are looking at the genre repertoire that has been used to standardize the humdrum and complex task of finding a reasonable newcomer to introduce into a workplace, and asking if the repertoire could be designed differently. Some want to add genres, such as Knack's measurements of qualities. Others, such as TalentCircles, want to change the means by which information between job seekers and employers is circulated in the first place. In all these instances, the companies are providing a platform that has built into its interface certain ways in which users will interact. It presupposes certain kinds of users and certain social relationships. While this is true of all things designed,[5] these companies are presupposing a user who resembles the self-as-business model more closely than older forms have. In all these cases, what the founders decided needs to be improved largely revolves around the ways in which older forms of hiring do not adequately reflect an understanding that the self is now a business. They are hoping with their designs to call forward a different business world, a business world in which the technologies people use to manage hiring more effectively reflect this new vision of how the world should be.

I am all too aware that describing the dreams built into a design is only half the story. Technologies are never introduced seamlessly into people's practices. Users always do unexpected and unlikely things with the technologies they start using. Because these were new companies, I rarely got a chance to find out what happened when actual users picked up these tools, which I suspect would reveal that they are not in fact the implied users that the designers had imagined. For example, I interviewed one recruiter who was using the programs designed by one of the companies that scoured the web for potential job candidates. He admitted that he didn't use the program to find candidates. Instead, he had discovered that it was extremely useful for uncovering possible hires' email addresses, addresses which companies often didn't make readily accessible for fear that recruiters like him would poach their talent. Since people tended not to respond to LinkedIn messages, he needed prospective candidates' email addresses, and it turns out that this program provided the addresses much more quickly than if he searched himself. Rather than using the program for what it was designed for, this recruiter used it as a glorified address book. You can build for a different world, but it requires an immense amount of coordination and labor on the part of countless others to use that technology in a way that actually changes people's interactions in the ways you had hoped, even a little bit.

This came to mind as another founder of a company was telling me that LinkedIn had managed to inspire a significant social change in how hiring was accomplished by getting people to post their resumes online for all to see.

> **Jim:** The biggest trick that LinkedIn ever pulled was convincing people that putting their resumes online had something to do with professional networking. At the end of the day, all the users who are on LinkedIn, they have zero concept that LinkedIn is selling their data out the back door to recruiters for ten thousand dollars a seat, on

LinkedIn Recruiter. But it was really impactful. . . . Historically, a resume, like when you're working on a resume and somebody came behind your laptop, you'd close your laptop and be like, "Hey don't look at my resume, that's my resume." Right? Whereas LinkedIn somehow changed that and made it such that it was A-OK for that information to be public, and it was not like braggartly. . . . And the other important thing was that like historically if your resume . . . if I'm a recruiter or a hiring manager at Google and I see one of my engineers' resumes in a resume database, like in Monster or CareerBuilder or whatever—

Ilana: It's signaling that they are about to leave. . . .

Jim: I'd get worried and then it would get back to the engineer and the engineer's like, "Oh, damn." So what that would actually do, was it would quash the willingness of these engineers, because they would think ahead, they'd be like, "Well, I'm mad at my boss, or whatever, so maybe I'll fill out my resume, but what if it gets back to them that I'm looking. Okay, screw it, I guess I won't." And so that would destroy information in the world. And so LinkedIn's big move was like, "No, no, no, it's professional networking."

As Jim talked about how transformative this was, changing how recruiters found candidates and altering what kind of information about people circulated readily, I thought about all the workshops I had attended teaching job seekers how to use LinkedIn. In every workshop, the instructor had to spend ten minutes or so reassuring various attendees that they did in fact want to be so public about this information. Government agencies were funding these programs in which people were told how to post their resumes publicly if they wanted a job. LinkedIn did not bring about this change on its own by creating the platform through which people could post their resumes online. Thousands upon thousands of people had to tell each other that this was now the new way to look for a job. And yet, as I showed earlier, people are still figuring out what LinkedIn is good for and how to use it.

In short, technologies do not call forth new worlds. People use

technologies to call forth new worlds, and in the process, they discover all the messy unpredictable ways in which actual users interact
with the complex social narratives built into the technologies they
use. What kind of world we all will create over the next decade may
be shaped by the technologies I have discussed in this chapter, but it
certainly won't be neatly defined by them.

Six   **The Decision Makers**

What It Means to Be a Hiring Manager, Recruiter, or HR Person

If you follow the self-as-business metaphor, then when you are being hired, you are entering into a business-to-business relationship with the company hiring you. In theory, you are deciding whether joining this company makes good business sense for your overarching career trajectory, and the company is deciding whether you can assist its operations. Yet hiring doesn't quite work this way in practice.

In previous chapters, I have discussed how complicated it can be for a person to represent him- or herself as a business, or as a human-shaped bundle of business solutions that can potentially address the market-specific challenges that a company is navigating. Now I want to shift to the perspectives of the people doing the hiring. As I quickly discovered, hiring decisions aren't often made based on market logic, and hiring is rarely a process in which everyone's role smoothly aligns. In chapter 2, I argued that workplaces are filled with distinctive, complicated social interactions, and the subtle political dynamics of one place are going to be different from others. Put another way, every company hires based on local circumstances and its own internal logic. At the same time, I heard over and over again about tensions in hiring processes that seemed reliably to emerge from the different institutional roles involved in hiring, not the company's internal dynamics. People most frequently talked about job applicants, hiring managers, recruiters, and HR. These institutional roles encourage people to come to different understandings of what makes a good hiring decision. For example, recruiters have many incentives to encourage others to make a hiring decision as quickly as possible. HR, by contrast, is often more concerned with making sure that hiring takes place as legally as possible. Hiring managers, meanwhile, are concerned with choosing job candidates who will be good fits (however that is defined) with their coworkers. These are all perfectly reasonable lenses for evaluating job candidates, but they are not always compatible. And while these vantage points are in dialogue with a market logic that seeks to create a profitable workplace, each suggests different paths toward that goal.

While these perspectives don't always fit easily with each other, people over the course of their working lives have often inhabited a range of these different, not-always-compatible perspectives. Everyone I talked to had been hired at some point in their lives, and almost everyone had participated in hiring other people, even if only as the administrative assistant scheduling the appointments. Most people had experienced at least two of these roles that everyone marked as important in the hiring process, and many of the recruiters I spoke to had experienced all four—having worked in HR, hired people for their own company, recruited for others, and periodically looked for a job.

This means that most people I interviewed had some experience with how your institutional role shapes how you participate and understand the hiring process. But even while people were trying to hire someone, they may not have paid much attention to the process. People often have a lot on their plate at work. Some roles might be more unfamiliar than others. If you have never worked as a recruiter, you might not know the constraints of a recruiter's job. When you are looking for a job, and recruiters contact you, how much do you know about the working conditions of recruiters and how that might be shaping why they are contacting you? As importantly for job seekers, how much do you know about why recruiters sometimes stop contacting you, seemingly out of the blue?

Even if job seekers had never been recruiters, they could still find out some information by attending free workshops. It turned out that, every now and then at the community-based organizations I frequented, recruiters would give presentations on how they read resumes, or how they use LinkedIn to find possible candidates. In doing this, they were talking about how they evaluate one or two of the many genres that job seekers have to produce to persuade others to hire them. Sometimes the workshops would consist of a panel of recruiters and HR professionals explaining what job seekers might need to know about their perspective on the hiring process. Recruiters outnumbered HR professionals though, which makes intuitive sense. Recruiters, after all, have a vested interest, both in meeting job seekers and in getting job seekers to present themselves in searchable and desirable ways so that the recruiter has appealing candidates to offer hiring managers. I have to say, self-interest wasn't the only motivation; the recruiters I saw were genuinely interested in being helpful to people who were unemployed.

Not many people in HR present at these workshops, and I never saw someone who had been a hiring manager come to talk about their process. The one time I saw someone from HR give a thirty-minute presentation, he did in fact give a very different presentation than recruiters did. While the recruiters tended to talk about how people should compose their resumes or LinkedIn profiles, or what to expect when interacting with a recruiter, the HR manager described what happened behind the scenes in the companies where he had worked. Vincent showed job seekers the evaluation forms he gives coworkers to fill out after an interview, and talked about the kinds of conversations people might have as they try to figure out whom to hire. This was the only time I heard anyone describe in these workshops the hiring process in terms of the forms coworkers fill out to evaluate interviewees or the debriefing meetings people have to choose a candidate. He was the only one who told job seekers that if there was a split decision among coworkers in a meeting where he had worked, then all five job candidates under discussion would be turned down, and the company would start looking at candidates all over again. He talked about the genres people use behind the scenes and about how these genres structure some of the ways that information around hiring circulates within companies.

While Vincent was an exception, for the most part, these workshops tended to focus on the interaction between two people: the recruiter / HR professional and the job seeker, or the hiring manager and the job seeker. If the focus was on the recruiters or human resources managers, then the discussion was all about how to deal with them as gatekeepers, as expert readers who were vetting resumes or interviews and always keeping in mind other people's requirements. These gatekeepers would describe themselves as having very specific styles of reading forms or evaluating initial interviews—for example, they claimed to spend anywhere from five to thirty seconds on a resume or LinkedIn profile. What they tried to explain to job seekers was how to compose the different parts of a job application to accommodate these ways of reading. When workshop leaders gave advice about interacting with hiring managers, a lot of the focus was on describing how to get ahold of hiring managers in the first place—how to get around the applicant-tracking systems and recruiters who might be keeping job applicants from what started to sound in these stories like an ever-elusive character. By the time I heard a few of these accounts, I started to get the impression that the hiring manager was The Decider.

When I started interviewing hiring managers, I realized how wrong this impression was. The hiring managers I spoke to described how many more people in the office contributed to making a hiring decision. They rarely saw themselves as The Decider, but rather as one of many involved in making hiring decisions that were caught up in the politics of their workplace. Yet repeatedly, the public conversations about hiring geared toward job seekers focused on how to have interactions with the hiring manager (and not the other people involved in selecting job candidates). So there was a mismatch between how job seekers understood what it meant to be a hiring manager and how hiring managers understood their role. Indeed, for the people in the role of hiring manager, being hiring manager was often only one of the roles they played over the course of a day. They sometimes only thought of themselves as the hiring manager because recruiters or HR would insist on labeling them the "hiring manager."

When I began interviewing hiring managers, recruiters, and HR people, it became clear that, for them, the interaction between everyone in the workplace weighed much more on their minds than their interactions with the actual job seekers. Indeed, occupying the roles of hiring manager, recruiter, and HR manager often put people into conflict with each other, largely because of the different institutional demands of these roles. Job-seeking advice tends to revolve around techniques for arranging job seekers' relationships with one role at a time, and often overlooks the social dynamics of a workplace, in part because it is difficult to describe these dynamics as standardized in the ways that resumes or interviews are thought of as being.

But, though workplaces are all unique, they do sometimes breed predictable conflicts, because there are so many moments in which doing your job well gets in the way of how someone else does their job. Anyone who has a job that involves guaranteeing that a business process follows certain regulations or is legally compliant has discovered that doing their job often prevents other people from doing theirs as quickly as possible. This problem doesn't always spring from enforcing regulations or standards. Stefan Timmermans describes how medical examiners often find that doing their job puts them at odds with all the other people whose job it is to deal with the recently dead. Paramedics will often bruise someone or break their ribs in their attempts to save the person, and in the process obliterate the signs forensic pathologists look for to determine cause of death. At the same time, the medical examiner often makes the morticians' jobs harder, as the autopsy the pathologist does to determine cause of death will disfigure the body in ways that morticians must then try to conceal for the funeral.[1] These kinds of tensions are present in workplaces where people with different jobs have responsibilities and functions that contradict each other in practice. These contradictions are a regular feature of any complex task that requires coordinating multiple people with different institutional perspectives. People frequently told me stories in which certain tensions kept emerging during the hiring process between recruiters, hiring managers, and HR professionals.

The roles in hiring, however, differ enough from Timmermans's account of forensic pathologists' clashes with their coworkers that it is clear that part of the tension results from the ways in which the work of hiring is allocated to different people. HR professionals can often do some of the work that a recruiter does (although a recruiter might suggest that they won't do it as well). Hiring managers can do some of the work that recruiters and HR do, if they only have the time. While in the case of autopsies, the jobs are so clearly defined that there is a clear boundary between what a mortician does and what a forensic pathologist does, and their tasks don't overlap, the same is not true for hiring.

Not every company parcels out the different responsibilities for hiring to the same institutional roles—for example, sometimes HR does work that in another company would be a recruiter's job. This is an attempt by companies to avoid the tensions I describe in this chapter. But when this work is reallocated, it means that these tensions are internalized. Perhaps a person given some of the tasks of a recruiter and some of the tasks of an HR person has to simultaneously encourage very fast decisions and decisions that adhere to formal regulations. When one person has to do more of these tasks, the work turns into the kinds of lived dilemmas I discussed in the introduction. As Jean Lave points out, "You *cannot* do everything at the same time." If someone constantly has to choose between different ways of dealing with the contradictory circumstances, they experience "dilemma management."[2] When, however, the task is stretched out over a number of people, then what in my previous example was experienced as one person's continually negotiated dilemma is more easily experienced as a personality conflict or a problem that could be solved if only people would communicate better. By making individuals take on only one horn of the dilemma, everyone involved finds it easier to overlook the fact that hiring always involves pulls in multiple directions, and that these are pulls that probably should not be responded to in the same way every time they come up.

All That Is Concealed under the Job Title of Recruiter

Often the first person whom a job applicant encounters in the hiring process is a recruiter. This is probably all that the applicant will know about the person contacting him or her—that the person is a recruiter. Yet knowing that someone is a recruiter does not in fact provide enough information about how a job applicant is likely to be treated. It matters quite a bit what kind of recruiter the person is. A recruiter's relationship to the company will determine how that person plays the recruiting game, and thus how he or she interacts with potential job candidates, hiring managers, and HR. While job seekers only mentioned to me that they had interacted with a recruiter, almost every recruiter I spoke to made sure early on in our conversation to distinguish between different types of recruiters and to explain to me which one they were. I interviewed various types of recruiters: sourcers, recruiters who work for staffing agencies, recruiters who have temporary contracts with companies to fill a certain number of positions, and recruiters who have yearlong or permanent contracts with larger companies. I even got to talk to a recruiter who worked for a venture capital firm. He had the task of developing a stable of talent whom he could persuade to join various start-up companies his firm had decided to fund, which meant he was playing the recruiting game slightly differently than all the other recruiters I met.

In what ways did a recruiter's relationship to the company affect the ways in which he or she selected candidates for a job? If a recruiter is an external recruiter, say part of a staffing agency, then it is likely that he or she is not the only one submitting job applicants to the hiring manager or to HR. A company can interact with a handful of staffing agencies, allowing each of them to propose possible job candidates for a position. The finder's fee will go to the staffing agency that finds the candidate who is eventually hired. This means that these recruiters don't have much or any contact with the hiring manager and won't know as much about what the hiring manager actually wants in a candidate. Their searches are mostly based on the job description. The recruiters are interested in finding as many candidates as possible who seem likely. It is a bit like throwing spaghetti at the wall and seeing which noodle sticks. At the same time, recruiters also have the incentive to look for people who aren't actively looking for jobs—the term of art for them is *passive job seekers*. If someone has already applied for the job, then the recruiter won't get the money. And if the person applies for the job through a friend or without letting the recruiter know, then the recruiter won't get the commission either. Often it is safer for the recruiter to approach people who are already in jobs and can be tempted to take another job. To deal with recruiters' preferences, one computer programmer told me that he would send his resume to every recruiter he knew when a job contract was coming to an end, and announce that he was an "active passive job seeker."

Why would a company or even an individual hiring manager turn to an external recruiter in the first place? William Finlay and James Coverdill explain in *Headhunters* that using a recruiter often turns out to be a solution to structural problems that a company or a hiring manager might face when hiring. First, companies often want to hire employees from their competitors or even their customers, but at the same time they don't want to antagonize the other company. This is a moment in which they might turn to a staffing agency or hire a recruiter on a short-term contract as cover to allow them to hire whomever they want without consequences. As Finlay and Coverdill point out: "Overt raiding of a company for its employees is a provocative action that invites retaliation against the offending organization." When companies use an outside recruiter, they can blame the recruiter for poaching the employee, using the fact that the recruiter is not part of the poaching company to evade responsibility for the recruiter's actions. The company employees might encourage the recruiter to find candidates from a rival or customer, since those are the employees most likely to have intimate knowledge of the market dynamics the company is navigating. The recruiter might even be told exactly which employee to approach to see if he or she might be interested in changing jobs. But as long as the recruiter is technically independent, the company can always use the recruiter as a scapegoat.[3] One company recruiter explained to me that she had toyed with this technique because her multinational company was moving into a new city where the only people actually qualified for her company's job openings already worked for the company's customers. She couldn't openly hire away the customers' employees, yet in that particular city, it was too hard to find good, skilled candidates or to get people to voluntarily move there. So she turned to external recruiters for a while.

Sometimes companies in the knowledge economy use recruiters as a different kind of cover. They are worried that job ads might reveal too much about their own business strategies or vulnerabilities. Whom a company needs to hire might indicate a new direction the company is going. Or the ad might indicate gaps in the company's workforce that need to be filled, but the company doesn't want to signal to its competitors that it doesn't have people with the essential skills. In these instances, recruiters won't post job ads. They will simply find likely candidates while flying under the radar.

Sometimes an external recruiter is hired not because a company is trying to address a structural problem but because a hiring manager is. Finlay and Coverdill point out that occasionally a hiring manager will turn to external recruiters to avoid dealing with HR: "Hiring managers lack the time and knowledge to engage in searches themselves, and they question how effectively the HR staff will conduct searches on their behalf. They feel that HR has its own priorities and cannot be expected to offer unconditional loyalty to any hiring manager. They prefer the unambiguous client-provider relationship they have with headhunters."[4] Recruiters, in short, are sometimes brought into the hiring process precisely because there is already a structural conflict in place between how hiring managers want the hiring process to go and how HR does.

The recruiters I have been describing for the most part work for larger staffing agencies, but not necessarily. Sometimes companies will hire recruiters on short-term contracts to staff their immediate hiring needs. These are smaller companies who believe that once they are fully staffed, they won't need the services of a recruiter anymore. Wayne, a contract recruiter, explained that people like him usually began in a staffing agency, matured as an internal recruiter in a company, and then started their own business: "If a job seeker is dealing with that individual, typically they are very senior, very Yoda like, they've been doing this a long time. They will be the candidate's best friend in the process, remembering this, recruiters want to make placement." For these contract recruiters, they succeed when they have made a hire that stays longer than sixty days. If they recruit someone who leaves after a short period of time, they have to work to replace that person, and they lose the money from the placement. It doesn't matter for them as much where the initial contact came from. Their salaries depend on getting people hired at the company who will stay.

These contract recruiters can serve some of the same structural purposes as staffing agency recruiters from the perspective of the company or the hiring manager. They can be the scapegoats in case a competitor or customer becomes angry that the company is stealing away employees, or they can run interference between the hiring manager and HR (these recruiters count as external enough for these purposes). But there are other ways in which contract recruiters have a different relationship with both hiring managers and HR once they have a more or less exclusive contract to recruit for a particular company, for whatever limited length of time.

For starters, these recruiters can often afford to be quite open about what company they are representing when they approach job seekers, unlike recruiters in staffing agencies. Recruiters without a company contract constantly risk being scooped by anyone else—including the job seekers who might apply without letting the recruiter know they plan to do so (perhaps because they have a friend in the company who would appreciate the employee referral reward). Recruiters who have a contract with the company are not so vulnerable. They are even in a position to encourage employee referrals, unlike other external recruiters. Thus, depending on what type of recruiter someone is, they are likely to share different kinds of information with promising job candidates.

There are, of course, recruiters who are on more permanent contracts with companies. Many large companies have in-house recruiters, and even sourcers (the recruiters whose sole job is to find likely candidates, but never to contact them). From the perspective of a contract recruiter, an in-house recruiter might often appear like an extension of the HR department. Wayne, whom I quoted earlier, was describing for me the lay of the land for recruiters and in the process differentiating his contract work from that of company recruiters. He explained:

> There are corporate recruiters who basically draw a salary whether they fill any position or not. Their focus is very much transactional and process oriented, and it is designed to make sure the company is compliant with the many number of hiring laws, both state and federal, that companies must adhere to today. It's almost a compliance role. It's almost a transactional administrative role. And many of these individuals have a strong HR background. They may or may not have the sales skills or really the recruiting background to truly be what I would consider a recruiter. Because at the end of the day, and most recruiters will hate it when I say this, but as a recruiter I'm a sales person. My job is to get you excited about an opportunity.

The corporate recruiters I interviewed would not necessarily describe their jobs in the same way Wayne did. They often talked about many of the challenges other recruiters report. They described the excitement of tracking people down who might be good, of figuring out how to find their actual email addresses or home phone numbers using any number of tricks. Companies, after all, protect themselves against having their employees poached by not making public any company email addresses or staff phone numbers. So recruiters would test the most typical patterns by which email addresses are created, or try to find an email address of any employee from that company on the web to see how that company tended to form email addresses. It wasn't always companies concealing who worked for them. Sometimes employees themselves did not want to be found. One corporate recruiter was particularly pleased that he had figured out that computer engineers of a particular type tend to misspell their expertise in obvious ways on their LinkedIn profiles so search engines can't find them easily. Once he figured this out, he was able to find these possible candidates by systematically checking a variety of misspellings.[5]

I spoke to one recruiter at a venture capital firm whose institutional constraints meant that he had a markedly different job than other recruiters I spoke to. For most recruiters, the job requisition defines their task. Various people at a company have agreed that they need to hire for a particular position, and the recruiter focuses his or her energy on filling the variety of job requisitions that are open at a given moment. Not so for Dylan—his venture capitalist company was interested in getting a large number of start-ups up and running successfully. As a result, he was never working toward filling a specific job. Instead he had a number of people whom he had vetted as talented workers who he imagined could flourish and transform any workplace. To find these people, he would repeatedly ask those he came across if they knew someone they would "work with anywhere." In looking for someone whom people would work with anywhere, Dylan was openly talking about one of the ideals of the self-as-business model—being a context-independent self. Once enough people named the same person, he would start courting. These were the people whom he kept in contact with over long periods of time. He would occasionally check in with them to see if they were content at their current jobs, or perhaps could be tempted into another position, one they saw as promising. In a sense, Dylan was working from an inverse starting point from most recruiters. Most recruiters will have a job requisition in hand and wonder whom they know who might be a good prospect for that job. The job description itself shapes how they approach searching and whom they will approach for the job. In a sense, Dylan begins with the people, finding candidates who seem to excel regardless of context. Then he tries to determine which company each person would find a tempting place to work from among the pool of start-up companies he helps staff.

The terms *hiring manager* and *HR professional* can in practice conceal the same complex set of obligations, incentives, and compromises that recruiters face, but these aren't as institutionalized as they are for a recruiter. The term *recruiter* can refer to a number of very different institutional relationships to hiring, which has implications for how recruiters with different relationships to a company are likely to circulate information and interact with hiring managers and job seekers.

Managing Humans as Resources

HR, while sometimes functioning as a sourcer or recruiter, often has another role altogether. HR can be the voice of law in an organization, watching carefully to ensure that hiring practices comply with state and federal regulations. One HR person explained this in no uncertain terms at the start of our conversation. "My philosophy, as an HR person, when I come into a company, is I represent the company. My job is to keep the company out of jail, out of court. So if something is going wrong," she explained, "and an executive is doing it, I'm going to be neutral. If it's an employee, I'm going to be neutral. I'm not going to take the side of an executive because they have an opinion. I'm going to take the side of the law." This perspective often ensures that HR managers will come into conflict with both the hiring manager and recruiter. Speaking for law, after all, often means that they are slowing down processes that recruiters want to have happen as quickly as possible and that hiring managers want to have happen as flexibly as possible.

HR's focus on law can put it at odds with recruiters or hiring managers in all sorts of ways. For example, as Brenda Berkelaar and

Patrice Buzzanell point out, HR and recruiters can have diametri-
cally opposed ideas about cybervetting because of their different
structural positions. The senior HR personnel whom Berkelaar and
Buzzanell interviewed were opposed to cybervetting, principally
because they felt that there were no consistent or standardized practices for finding online information about job applicants. The information you found out about one candidate could be wildly different
than what you discovered about other applicants, and those hiring
rarely followed the same procedures for gleaning information for
every candidate. They wanted "systematized and legally compliant
processes, codified in professional ethics guidelines . . . and likely
reinforced by their occupational identification." This was a very different attitude than recruiters had toward cybervetting. Recruiters
wanted to find possible candidates any way they could and weren't
concerned about searching in standard or legally compliant ways.[6]

As the one overseeing how the hiring process is being run, HR
is all too often depicted as the central obstacle by others within the
company. While job seekers might talk about getting around the
applicant-tracking system with the same determination and exas-
peration as they talk about getting around HR, this isn't the same
on the hiring side. Instead, HR alone often looms large for hiring
managers and recruiters as the primary, and all too human, obstacle
as they try to hire.

While HR people are concerned that hiring is done legally, they
often try at the same time to accomplish another part of the job, one
which is not always compatible with being the voice of law in an
organization. HR professionals will also try to get their coworkers
to think consciously and strategically about their hiring. People in a
company are often so busy with their daily tasks that they don't have
a chance to think strategically about where the company is going and
what kinds of employees it will need in the future. Peter Cappelli
points out that this type of foresight is a role that HR can take on by
anticipating future staffing needs.[7] In my interviews, I came across
various instances in which HR people were trying to do this, not
always to the hiring manager's pleasure. John was working at a small

company as its accountant and wanted to hire someone to manage day-to-day billing. John needed someone quickly. The last person had quit for medical reasons, and his department was strapped. But Leila, the HR person, wanted to think more expansively of what the job could be, and she started explaining to the CEO that this was a job whose description should be written much more broadly. John was frustrated. He needed a person quickly. But now he was faced with office politics in which he had to contradict a coworker, largely because anyone who applied based on the job description Leila wrote would be disappointed to find out what the actual job involved. In this instance, HR and the hiring manager were forced to balance the company's immediate needs with an attempt to antici-pate future demands.

In short, one of the roles HR can fill is to provide strategic ori-entation to hiring decisions so that hiring is understood not only in terms of the immediate needs of the company. That is, HR can ensure employees are seen as potential resources for the future, so that a company hires a bundle of skills who can transform him- or herself in a year or two for the future needs of the company. HR pro-fessionals thus often have to balance regulations with future strate-gies, choosing which they want to act on in a situation.

While not every HR professional views him- or herself as a strat-egist for the future, HR professionals do invariably see themselves as organizing the nitty-gritty details of hiring. And in the process of organizing hiring, they are often asking people to explicitly reflect on why they are choosing one candidate over another. HR is part of the human infrastructure that makes hiring happen, and thus it often structures the ways in which people judge candidates. When a team is hiring as a whole, the HR manager may be both outside of the team and yet part of the company, having a privileged vantage point for seeing how the team has developed its own ways of doing things in the context of the larger company's shared sensibilities. In these moments, HR is often moving rapidly between being the voice of law and the voice of general, shared company expectations—in each instance, speaking as a bit of an outsider to those making the final decisions about who will be the next newcomer to the company. HR, in short, is constantly balancing the need to act according to the law, the need to act according to the strategic potential of the company as embodied in the company's employees, and the need to make sure that the processes for hiring are effectively in place.

One Obligation among Many

Hiring managers, meanwhile, find themselves often balancing a different but related set of dilemmas. The first issue that came up whenever I talked to people about being a hiring manager or working with hiring managers was how difficult it was for people to find the time for hiring. Vincent, the HR person I mentioned earlier, might say that this was simply because people in general hate hiring so much. He thought it had more to do with people's personalities. I told him that hiring managers just kept telling me they didn't have enough time to devote to hiring. He disagreed: "They hate it! They just hate meeting others. Can you imagine an introvert scientist that says, 'Now I've gotta go interview someone and hire them for my team. I've gotta go meet people'? I've had to explain, 'Well, Cinderella. You kiss a lot of frogs. You've gotta go find your prince!'"

I heard a number of other reasons why people hate the task of hiring. They hate having to reject people for a job, being cast in the role of evaluating whether people will join a workplace, and making this decision based on what they feel is too little information. The work of hiring often takes you away from demands that seem more pressing, while hiring is the unpleasant task that has too much ambiguity and too much office politics at its core. To find someone to hire, you are often forced to deal with anxious people who are trying very hard to conceal how nervous they are. You are also forced to deal with other people's skills at evaluating candidates within a relatively limited amount of time, including your direct supervisor's skill at evaluating someone. At the same time, you are often hiring because you need someone to do particular tasks, tasks that are currently being distributed among other people at the company who feel overwhelmed and overworked. And actually hiring someone does not immediately solve the problem, because you have to take time to train the person, to help them adjust to the idiosyncrasies of your work and your workmates. While hiring someone will often eventually ease the time pressure people are working under, this is not always an immediate solution. And so hiring may be in practice the task that those in charge of the hiring decisions are most likely to want to avoid.

Hiring is also a particularly loaded activity because it is the moment in which people in an office are most openly reflecting on what those I interviewed have come to call a company's culture. When they are choosing someone to hire, they will frequently talk about cultural fit,[8] and determining cultural fit often involves talking openly about the ways people think that everyone at a company interacts (whether or not this is in fact the way they interact). Hiring is a moment in which people are most consciously trying to address the question of who they are as a working community in order to understand whether an individual will fit. So choosing someone is in a sense announcing to others who you think the group is and what kind of person belongs in it. This is often a loaded negotiation, with those hiring constantly trying to anticipate other people's perceptions of what the company is like and who will fit in well in that particular work environment.

When Roles Clash

The tensions are all fairly predictable. Hiring managers rarely have time to do justice to the complexity of what it means to hire. They are being pulled in a million different directions, often in part because they are trying to compensate somehow for the fact that they need a worker whom they do not yet have on the team. When recruiters or HR managers reflected on their relationships with hiring managers, they often described how difficult it is to get the hiring manager to pay attention to the hiring process until the interviews begin. They described in detail their strategies for getting hiring managers to focus or to respond to their concerns, strategies that often involved manufacturing focused conversations around documents. Recruiters and HR people repeatedly presented this as a problem of inattention which could be solved with more communication. I have my doubts—these sound like structural problems to me.

Recruiters meanwhile are largely focused on hiring people as quickly as possible. All recruiters have strong incentives to fill job openings, although recruiters who are in-house at companies may face the least amount of financial pressure to do so. Hiring managers want to hire the best person they can find, or at least a competent, socially unproblematic, and reliable person—the good enough candidate (if they are thinking about this dilemma as a manager, and not as a consumer). Recruiters want to place a reasonable candidate as quickly as possible, and what counts as reasonable for them largely depends on whom they can persuade the hiring manager to hire. And HR wants the hiring process to be legal. No wonder there was bound to be some tension between hiring managers, HR, and recruiters.

When More Communication Does Not Solve Communication Problems

While people often talk about these tensions as personality issues, they are misunderstanding the dynamic that is in fact taking place. I so often heard the same types of frustrations expressed about people who clearly did not share the same types of personalities that it became apparent over time that many exasperating hiccups in the hiring process are caused by people's institutional roles, not the personal qualities someone has. While everyone finds some way to work around the inevitable tensions that come with the roles, all too often people identify a person's personality, not his or her job, as the cause of the headaches they experience. And when these issues emerge, they often surface around how people engage with the documents that are central to the hiring process.

For instance, composing the job description is often the first moment in which it starts to become apparent that hiring managers have obligations that will make it harder for recruiters or HR professionals to do their job. The job description is a crucial text in the hiring process, and not only because, when circulated properly, it announces to potential applicants that a job is available. It is the primary text that job seekers will use to craft their application. They will try to determine what the keywords are in the job description, and then pepper their own application with these exact words to get past the application-tracking systems. Some applicants will even compare word clouds of both the job description and their job application—that is, they will create a graphic representation of the word frequency in the job description and check that their job application contains the most commonly used words in the job ad. This is the main text that job applicants interpret to figure out how exactly they can best tailor their application so that they seem like a bundle of business solutions specially suited to this particular business's needs. Yet time and time again, HR professionals told me that they got the text for a job description by looking online for similar-sounding jobs. In short, the text that job seekers interpret as specific to a particular job and a particular company is often first written for a different job at another company.

This happens because HR or a recruiter is often given the task of composing the job ad, which is then sent to the hiring manager for approval. To do this, the HR person or the recruiter has to translate. They don't do the job being described. They don't work on the team. So they have to take what information the hiring manager has provided about what the job involves and translate this into a description that is comprehensible and appealing to people who know how to do this job. Hiring managers sometimes know exactly what they want in a way that they can explain effectively to recruiters or people in HR. But often they don't. Faced with an incomplete explanation, the recruiter or HR person will turn to online job descriptions to see models of how others have composed ads with this job title at similar companies. Hiring managers will look at the job description, but they often don't have time to edit it carefully. So they will glance at the ad. If it looks good enough for what they are vaguely hoping to find, they approve it. While this seems benign, this quick approval can set the stage for future complications and misunderstandings. This is the first moment in which the hiring manager and the recruiter or HR person could have a careful discussion of what precisely is needed for this job, but as my interviewees told me repeatedly, in their experience this rarely happens.

Of course, job descriptions themselves are a strange genre, so perhaps it is asking too much to expect any conversation to lead to an effective written encapsulation of what a job involves. Job descriptions presume that the complicated social dynamics involved in doing a job alongside other people can be reduced to a list of clearly delineated skills and tasks. As Jean Lave points out, job descriptions are written "as if 'the job' were a limited series of tasks requiring specific knowledge and skill, tasks that can be easily listed. They also assume that everything the job doer needs to know is taken care of when the individual has mastered those tasks."[9] Yet even analyzing the work that goes into hiring new people for an organization reveals that no list of tasks captures adequately the complex social interactions you have to navigate in order to choose one applicant out of a pile of many possibilities. Job descriptions often function to keep people from explaining to each other what they actually want in a job candidate, because creating a job description means turning to standardized ways of talking about what might be required in a job, instead of capturing what the daily life of being part of that particular workplace might involve.

Internet job boards have only served to exacerbate this problem, as sociologists Emmanuelle Marchal, Kevin Mellet, and Geraldine Rieucau found in their comparative study of internet job boards and newspaper job ads. They were interested in figuring out what effect job boards had on the way information circulated between recruiters and job seekers. Job boards, after all, are in theory a neutral technology, geared toward helping both parties in a search find the best match. The job board should be structured to circulate information equally well to everyone involved. And yet, Marchal and her colleagues found that internet job boards, in order to enable keyword searches, forced job descriptions to include standardized terminology that often didn't reflect what those hiring were actually looking for. The emphasis on classificatory schemas also prevented job seekers from finding the best fits for their distinctive collection of skills and experiences. This problem only intensified for people who could apply to a wide range of jobs because their skills were (ironically) too transferable. It also was a problem for employers searching for a set of skills or job roles that were so new that there were no industry-wide terms to describe the jobs.[10]

Yet while job descriptions are a standardized form in which the very standardization they embody can create communication problems in the hiring process, recruiters might use another standardized form, resumes, to overcome potential communication problems. This is not possible for recruiters in staffing agencies, since recruiters at staffing agencies rarely have access to a hiring manager. Instead, any number of agencies might be submitting possible job candidates to the HR person, who then has to sort through the possibilities based on what he or she thinks the hiring manager wants. The recruiters who work one-on-one with a hiring manager, say those who work for the company on some kind of contract, have different options. These recruiters often told me that they have trouble figuring out what the hiring manager actually wants in a candidate. The first conversation or two might not be all that helpful. They deal with this problem by collecting ten or fifteen resumes and showing them to the hiring manager. They watch as the hiring manager reads the resumes, and ask why he or she likes some candidates and not others. In this way, recruiters try to get hiring managers to talk not in abstractions about what they would ideally like to have, but to focus on the merits and failings of the actual candidates. This is how some recruiters are able to get a sense of the kinds of practical compromises a hiring manager might be willing to make in the future. But even these recruiters often felt frustrated.

At first glance, you might think that hiring processes should typically go smoothly. After all, hiring is essential for a company to survive—all communities need mechanisms for adding community-appropriate newcomers so that they can maintain themselves over time. And, for the most part, everyone involved in the hiring process wants to hire. In all the searches I know about, I never heard a story about a problematic or failed search where the issue was that people simply didn't want to hire. The problems always lay elsewhere. People may not have had the time to hire, or may have disagreed about whom to hire, but no one seemed to believe that hiring in the first place was a terrible idea. So everyone involved wanted the process to end with a hire, and in order for a business to function properly, there has to be hiring. Yet, as I noted throughout this chapter, people's roles often clashed with each other enough that there were serious problems in the hiring process. Even if people were performing their jobs well, this still could lead to a dysfunctional hiring process. On paper, this might seem unlikely. Why shouldn't it be possible to smoothly and easily coordinate one person who searches for job applicants, one person who makes sure the search and decision are done properly and according to regulations, and one person who makes the decisions? And yet in practice, time and time again, people found that doing their jobs meant keeping other people from doing theirs as smoothly as they would like.

| Seven | **When Moving On Is the New Normal** |

Quitting (and getting laid off or fired) is the engine that makes contemporary hiring what it is today. Recruiters are constantly trying to figure out how to get people to quit (for another job), and many people talked to me matter-of-factly about someday quitting their jobs as the obvious next career step. After all, if every job is temporary, and a career is really a string of jobs, then quitting (or getting fired or laid off) is always just around the corner. This is yet another aspect of how the self-as-business metaphor dominates people's strategies in the workplace—a job is now a connection with another business which ideally enables you to enhance your desirability to many other potential employers. Everyone "should" always be anticipating their next job.[1]

Quitting doesn't mean what it used to. A common view has been that quitting is typically a strong critique of the organization or community that you are no longer willing to be a part of. The economist Albert Hirschman talked about this in *Exit, Voice, and Loyalty*. Exit for Hirschman is one response people can have when they want to critique or repair faltering institutions. It is a strategy people can use to point out that the ways in which a workplace is organized create problems that could be fixed with some effort, but, for whatever reason, the person quitting is no longer ready to put in the effort. This idea lies behind the popular stories in which someone dramatically resigns, loudly proclaiming their dissatisfaction in such an unequivocal way that they can tell their quitting story for years afterward. Yet using quitting as critique comes at a very high cost in today's job market, as it risks cutting off ties. Nowadays, for many people, the challenge of quitting is figuring out how to exit while maintaining as many relationships as possible.

The rest of the book has taken hiring as the starting point. But if you take quitting as the starting point, then you pay attention to how job candidates nowadays make decisions about their current job search in terms of how the job they are considering taking will help them get the next job. Being promoted within the company is still something to consider, but many now believe it shouldn't be the main focus. In Silicon Valley, this is even true for the most prestigious companies where someone could get a job. People talk about wanting a job at Google or Facebook because it would be so much easier to get the next job if you were already at a well-respected company.

How does this work in practice? Rose's first job after graduating from college was as a communications manager in a nonprofit organization, a position, she explained, she had felt stuck in—everyone there was constantly working too hard, and the expectations felt too unrealistic. In fact, she was so overworked that she had very little time to apply for other jobs. She started planning her exit strategy, but she wasn't sure what job she wanted next. Finding a similar job was not going to be an improvement. She would still be facing all the same problems, just at a different organization. It became clear as I talked to her that she decided to think about the job search in a new way.

When she took the job she eventually wanted to quit, perhaps the only thing that she knew was that she wanted a job in a nonprofit organization, but she was willing to take any job after graduation that she could get. In hindsight, she was beginning to realize that she had to have much more sophisticated criteria. She began asking a select few people for advice, and she forwarded a job ad that appealed to her to a VP at another nonprofit organization whom she happened to know. He pointed out to her that the job ad was for something specific to her state, involving that state's legal regulations. If she took that job, she might not be able to move as easily out of state. He explained that she was not just deciding if she wanted *that* job, but she was also deciding if she wanted a job that would more permanently anchor her in that state. He also thought that she should try to move into a field where there were many nonprofits, so that she wasn't limiting herself to a very few companies when she decide to move organizations again. Rose said he taught her that finding out the typical trajectory of people who had worked at the company was also important: "He gave me some thoughts about making sure the organization is healthy because you can tell a lot from how long the staff has been there, what the turnover is. . . . He told me to look up the person who had the job before you, if you can find out, and seeing if that person went on to a higher job in another organization. Does that mean that this job gave them skills to then become a director level, you know, or are they getting promoted up within the organization?" What the job itself might be wasn't the point. Instead he urged her to weigh more seriously what the job might lead to in the future.

While people have long judged the jobs that they might take in terms of future job opportunities, they also used to think about the possibilities of mobility within their company. Even though moving up within a company is still something to consider, it is no longer people's primary focus. What is distinctive nowadays is that mobility largely involves planning to quit the job you are currently considering accepting for a job at a yet unknown company.

Managers also recognize this. Being a good manager is now understood as helping those working for you develop the skills that will enable them to find another job. I attended one workshop for new managers in which people who had been recently promoted to a managerial position were taught how to handle their new role. One speaker modeled for everyone in the workshop how she supported her employees by taking them out to lunch in the first week: "So I always say things like 'You don't work for me, I work for you.' . . . 'My job is to make sure you can do your job well. And one day, you are going to leave this job, right, our careers are long, and we will have many jobs along the way. When you want to leave this job, I hope to be here to help you move on to this next job.'" This speaker described framing her relationships with new hires in terms of their eventual decision to move elsewhere. In short, from the beginning, both employees and their managers are openly talking about jobs as stepping-stones.

This has changed the nature of work in some companies, as employees try to enhance skills that will make them attractive for that nebulous next company and avoid tasks that seem too specific to the company they currently work for. Because my own fieldwork was focused on how hiring takes place, I never spent time in workplaces watching workers and managers wrangle over the division of labor. Linus Huang, however, is a sociologist at Berkeley who was doing fieldwork in a Silicon Valley start-up company just as Java became a popular programming language.[2] Most of the programmers at the company he was observing quickly determined that mastering this computer language was likely to help them find their next job. At the same time, many of the software company's immediate needs did not require programming in Java, as C++ was more than sufficient for most projects. Managers often needed programmers to work on certain tasks that were specific to that particular company's needs and linked closely to the company's intellectual property. Working on these projects meant spending weeks accruing skills and knowledge that the company would not allow you to use elsewhere. And this, as I mentioned in the introduction regarding noncompete clauses in employment contracts, is a problem that is starting to spread beyond software companies. That is to say, from the workers' perspective, not all the tasks were equally beneficial to their self-interest, assuming they wanted to enhance skills that would provide them with greater job opportunities outside the company. As a result, people in Huang's office were constantly jockeying to be assigned certain projects and not others.

Managers found this situation challenging to negotiate, since often the tasks programmers wanted to take on were not the most urgent tasks from the company's perspective. Managers had to keep trying to allocate projects in ways that undercut workers' self-interests. They often turned to the few programmers who had not yet decided to plan their workdays around quitting at some point in the future (and these often were some of the most senior employees at the company as well). A tacit division of labor developed in which people who fully took on the self-as-business model did some kinds of tasks and the few who thought differently got other kinds of assignments.

Not all tasks are created equal when workers are constantly anticipating having to market their skills to other companies. Some tasks are too contextually specific, too bound up with a company's intellectual property or too linked to the idiosyncrasies of that company's products. Thus the company's interests and the workers' interests don't always align when managers have to distribute some kinds of technical work, which can cause tensions in the workplace.[3]

Becoming a Passionate Worker

Now that quitting is such a central part of people's decisions when they take a job, companies have a new dilemma. After World War II, businesses had a problem retaining skilled workers, or, indeed, workers in general. The economy was booming, and workers could demand higher and higher wages, especially with strong union support. Employers were faced with a problem: either find a way to keep workers at their companies or risk entering into a potentially costly competition for labor. It was at this historical moment that companies decided to promote the concept of company loyalty. Company loyalty fulfilled certain social functions for a corporation, including ensuring that it had a reliable pool of talent that was already familiar with how the corporation worked. In addition, the company had people who could serve as repositories of knowledge—who not only understood the intricacies of how various parts of the company worked, but could remember earlier efforts at responding to regularly occurring problems, with insight into why certain solutions succeeded and others failed. Businesses were operating with an eye to the future, and they would have a guaranteed supply of properly trained workers promoted from within their own ranks, understood to be properly trained because the companies had trained the workers themselves. They might have had to invest in training, but they tended to reap the benefits of this training. Teaching a worker how to do certain jobs was not putting effort and resources into someone who would soon leave for another organization.

While there are many reasons to want a stable workforce if you are thinking of the long-term well-being of a company, it is not a financially viable goal if you are thinking about the short-term fiscal health of a company. Karen Ho, in her ethnography of investment bankers, *Liquidated*, discusses how over the past thirty years, corporations began measuring a business's financial success in terms of how it was doing each quarter, instead of taking a long-term focus. She argues that in the mid-twentieth century, corporations believed that shareholder value depended on the ways in which a company contributed to stable careers and stable communities. Since then, corporations have changed their philosophies—their present concern is with keeping their stock prices as high as possible. Ho argues that this shift happened because corporations changed their understanding of who contributes to making a company what it is and whose interest a company serves. At the time that company loyalty was seen as an important value for keeping workers engaged and committed to companies, businesses were seen as the product of many different constituents and were believed to be responsible to all these constituents. Businesses were successful because of all the employees' hard work, as well as the guidance of the founder, subsequent owners, and corporate boards of directors. As a consequence, business plans should keep their interests in mind as well. Yet Ho points out that over time a different model came to dominate, one which claimed that all decisions needed to revolve around the shareholders' interests, and the shareholders' interests were defined only in terms of the stock price. This economic model meant that people would make decisions that would elevate the business's stock prices in the short term, say through mergers, despite the fact that time and time again such strategies had harmed the company's abilities to be profitable over the long term.[4]

One of the most common decisions companies began to make to seem profitable in the short term was to downsize their workforce, outsource jobs, and put as many of their employees on temporary contracts as possible. Peter Cappelli argues that this began to seem like the wisest human resource management strategy after a combination of historical factors. Having spent decades building a strong internal base of workers, in the early 1980s, companies had an excess of trained workers; they had overestimated how many skilled employees they would need. After the 1981 recession, employers decided that they needed to rethink the implicit promises they had made to employees to offer both a career and a job. They began to urge employees to stop thinking of themselves as permanent employees. While this concept took awhile to spread across industries, it was very much in place by the late 1990s. A Conference Board survey in 1997 revealed that only 3 percent of businesses provided their employees with job security.[5]

Employees began to notice that without certain company incentives in place, they needed to change their career strategies. While previously workers had found that changing jobs too often meant their salary wouldn't rise as quickly, this began to change in the late 1980s and early 1990s. Economist Dave Marcotte found that men's salaries had begun to rise at the same rate whether they stayed in a job or switched companies regularly.[6] By the twenty-first century, pensions were rare and seniority at a company did not go hand-in-hand with a higher salary. Most employees no longer pay any financial penalty for switching jobs.

This was a calculated change in tactics for businesses. Vera worked for decades at a community organization for job seekers in Silicon Valley, and she told me about the moment when companies began to approach her organization to help explain to their employees that the logic of the workplace had changed. In the mid-1980s, senior-level HR people contacted her, saying "We want you to come in and teach self-reliance because we can't promise any longer what we once did in order to stay competitive." When companies moved toward valuing shareholders and short-term profits, they had to rethink the promises that they made by encouraging workers to value company loyalty. Workers got the message, although some are still nostalgic for the long-gone job stability. As Carrie Lane found in her fieldwork with unemployed high-tech workers in Dallas, "whether suspicious of or nostalgic for that bygone era, nearly all agreed that the days of corporate loyalty are over. Long gone, and likely for good."[7]

If company loyalty is no longer the promise that encourages workers to come to work and pay attention while they are there, what is the emotional connection that keeps workers willing to work at others' direction? There are many options. Guilt, for one, works particularly well on me. Tell me convincingly that I am under obligation to someone, and I will stay up late at night. But this is not the emotion that US employers and many self-help books turned to in their efforts to persuade workers to behave in ways that employers want. The emotion instead is *passion*.

I heard about passion all the time. What the job actually is, and whether the job does in fact require that someone feels strongly about their work in order to do it well, seems irrelevant. Passion has become such a frequently repeated word that everyone involved in hiring seems to agree that one of the most important qualifications for someone to show in an interview is an overwhelming enthusiasm for the job. A manager explained to me that he would much prefer choosing the not-so-talented person to work on his project as long as he or she was passionate about the work. Passion for him trumped skill, trumped experience. When I asked him why, he laughed and said that this is what guarantees that the employee will work the long hours necessary to get the job done. He could teach the skills to someone who didn't already know how to do a task, but he couldn't make someone deeply committed to tasks if they didn't feel committed from the outset. Indeed, the only time people never talked about passion during my fieldwork was when someone described hiring for contract jobs or temporary jobs—jobs in which people often couldn't work extra time even if they wanted to, or else they would perhaps cost the company overtime pay if they did feel such a passion for their work that they didn't want to stop. But if a permanent job was on offer, however short term that "permanent" job might be in reality, then passion was the key emotion.

Just as an employment system that required company loyalty inspired deep ambivalence among companies and job seekers in the 1970s and 1980s, an employment system that requires passion has its downsides too both for employers and job seekers. In his book *Flawed System / Flawed Self*, Ofer Sharone argues that requiring job seekers to express passion in job interviews has detrimental effects for American job seekers.[8] He points out that many self-help books these days argue that you should only work on tasks and in jobs that you feel passionate about. This of course often leaves job seekers at a bit of a loss. The things that they might feel passionate about are not necessarily lucrative. You may love gardening, or raising your children, or going to church, but these are not necessarily what you want to focus on for work, although Sharone does point out that part of the passion logic involves insisting that everything you might like to do can be monetized and should be if that is what you feel passionate about. By insisting on the financial potential of everyone's passions, the self-help advisors who are advocating passion are also advocating ignoring inequalities among occupations. Feeling passion for gardening may not lead to a well-paid job, while feeling passion for merging companies might.

But more than this, Sharone believes that the language of passion leads white-collar job seekers to blame themselves when they can't find a job, rather than taking into account larger structural or economic reasons. After all, if you can't get a job, perhaps it is because you haven't convinced other people that you are passionate enough about that job. He found that Americans frequently blamed themselves and their inability to be personally persuasive enough when they had trouble finding a job, not the lack of adequate jobs, corporate restructuring, or systemic discrimination of various kinds. Sharone argues that focusing on passion keeps people from framing their unemployment in terms of systemic problems that require systemic solutions. Instead, the only solution they are able to imagine for their situation is to figure out what they are in fact passionate about or to try to find techniques to be more persuasively passionate in interviews. Motivational speakers and self-help gurus repeatedly tell people that following their passion will inevitably over time lead to success, regardless of the obstacles. From this perspective, if it did not work, then the flaw is an internal one—the job seeker wasn't passionate enough.

While job seekers may be made more vulnerable by this emphasis on passion, companies become vulnerable as well. When people talk about feeling passion for their work, they are talking about feeling passion for a certain set of tasks or learning a set of skills. They don't talk about feeling passion for the ways in which their coworkers interact or a happy workplace. They also don't talk about feeling passion for making the company they work for well-respected among other companies. Passion is reserved for the tasks they do or learn to do, for the solutions they might develop for market-specific problems that the company faces. And all too often, the market-specific problems employees talk about being captivated by are problems a range of companies might face. They aren't specific to that particular company. In short, working based on passion is focused on the task, not the company. And that makes workers more mobile. They can more easily contemplate moving to a new company. One executive recruiter told me she used this focus on passion to help persuade executives to leave, regardless of the financial incentives put in place by their current company. The passion logic could unlock golden handcuffs. She would explain to executives that if someone no longer felt any passion for their work, then they were harming everyone involved, including their coworkers who had to work with someone no longer wildly enthusiastic about the job. When the main reason to work somewhere is because you feel passion, it is all too easy to quit because you have stopped feeling passion. And while companies no longer want to encourage the long-term obligations to employees that come with inspiring company loyalty, companies sometimes suffer from the ways they lose employees because passion is the emotional touchstone.

At the same time, some companies have started to respond to the ways in which passion and quitting are now intertwined by offering people financial incentives for quitting if they are ambivalent about or unhappy at their jobs. In 2008, Zappos began offering cash for quitting, and Amazon followed suit in 2014, offering a couple thousand dollars to people if they quit their jobs. The companies announced that they didn't want employees staying who lacked enthusiasm for the job, and they were happy to provide a financial incentive to encourage people to think carefully about whether they wanted to stay in their jobs. In 2014, Amazon's CEO circulated a letter to its shareholders that explained the logic of its version of the program, Pay to Quit:

> Once a year, we offer to pay our associates to quit. The first year the offer is made, it's for $2,000. Then it goes up one thousand dollars a year until it reaches $5,000. The headline on the offer is "Please Don't Take This Offer." We hope they don't take the offer; we want them to stay. Why do we make this offer? The goal is to encourage folks to take a moment and think about what they really want. In the long-run, an employee staying somewhere they don't want to be isn't healthy for the employee or the company.

Amazon is in a sense asking employees once a year to reconsider their employment contract, to decide all over again that this really is the job they want to keep. Paying employees to quit makes sense in a context where companies look for ways to guarantee that passion is the emotion that keeps employees committed to their jobs, just as paying people a pension made sense in a context where companies sought to ensure employee commitment through loyalty. This is but one example of how companies are currently experimenting with a new set of incentives to encourage employees to be passionate instead of loyal.

In the media, following your passion is a highly valued reason for quitting, especially when quitting then becomes the basis for a career transition. Most of the media stories about quitting that I have collected revolve around how someone discovers their passion, which then leads them to quit their job and turn to an unstable, financially risky, and yet ultimately rewarding career. What are the common elements of this story? Someone decides that they are stuck in an unfulfilling job. They figure out what they would really love to do instead. (In reality, the job seekers I spoke to often became stalled by the complicated and too loaded challenge of finding their passion.) They then find ways to develop skills in this area through some combination of taking on new tasks at their current company, working two jobs, taking classes, or volunteering. Then, one day, they decide to take the plunge and quit their job to work in this new, often freelance, job that becomes remarkably lucrative because, of course, that is what happens when you follow your passion.

This has become such a commonplace story that I was a bit surprised to come across an equally personalized *Salon* story warning readers that quitting to follow your passion can be a financially disastrous path to take. David Sobel writes about how listening to a talk about following his passion inspired him. He decided to quit his job at a policy organization and start working as a digital copywriter at the age of forty-two. Before he did this, he tried to improve his digital marketing skills at his company, volunteering to do all sorts of tasks that could signal to potential employers that he was now proficient at this new line of work. He took an advertising class. And finally, he quit his job, determined to put all his energy into searching for a job in his new line of work. He found nothing. He was faced with the familiar dilemma I have written about in earlier chapters: no one wanted to hire someone who was middle-aged and without enough experience. For those hiring, a lack of experience is often excusable only in recent college graduates, not those who have left behind one career to attempt the so highly praised decision to do what you love. He concludes his story by explaining how he got fired as a dog walker and couldn't get hired at Starbucks. Sobel's story suggests that thinking in terms of one's inner passions not only can lead people to blame themselves instead of structural inequalities, as Sharone suggests. Thinking in terms of passion also encourages people to overlook the larger structural constraints on their decisions, whether as job seekers or as quitters about to become job seekers. Indeed, that is the whole point of following your passion.[9]

In actuality, when you quit to follow your passion, you might be crossing into social contexts that reveal that not all the infrastructures or everyday assumptions about careers have been rewritten to accommodate this new vision of work.[10] This became clear to me as I talked to Sucham, who had left Google to develop a catering business based on her parents' longstanding restaurant. Sucham had gotten a master's degree and immediately been hired at Google to encourage businesses to advertise with Google. She worked there for four years and found herself increasingly unsatisfied. Working at Google taught her quite a bit about how business operated, but she began to realize that what she wanted to do was not sell advertising but sell food. Her parents had owned a restaurant for years, and she loved their recipes. She thought that with what she learned at Google, she had the know-how to expand their restaurant business and start catering around the Bay Area. Eventually she could open another branch or two of their restaurant.

When she quit her job at Google, she began to cater some of their events—a contact through a local food truck organization helped make this possible. And that was when she found out that working at Google and working for Google were two different things entirely. She said that people she met while catering events at Google often did not understand when she said that she used to work at Google. They would ask her if she had worked preparing food in the cafeterias. The idea that a marketer would choose to become a caterer seemed foreign to them, despite all the advice that circulates suggesting that everyone should follow their passion.

Sucham also found that having worked at Google did not give her special privileges when she catered events at Google, although I admit that *special privileges* does not seem quite the right term to describe the following story. At one point while she was catering an event at Google, Sucham became thirsty and wanted to get a bottle of water. She was used to opening up the refrigerators that are in almost every Google office and helping herself to a water bottle stored there. But before doing this automatically, she paused and thought to ask the Google employee whom she was working next to if she could grab a bottle of water, if the employee thought that this was acceptable. The answer was no. The employee was very clear that the water was only for people at Google. Sucham told me this story because it was emblematic for her of how her status had changed: following her passion had demoted her in social hierarchies in a tangible sense. While Sucham's narrative of quitting her job to follow her passion fits well into one of the more idealized narrative arcs about how quitting is supposed to take place, it also points to the ways in which not all passions have equal outcomes. If your passion is to figure out how to make money on the stock market, following it will yield different social and structural consequences than if your passion is to build beautiful houses or to take care of other people's physical or psychological needs.

The Work Involved in Quitting

What these quitting stories don't reveal is how difficult it can be to get to the point where you can quit your job for a new job or to take on a new project. Jobs you want to quit are often jobs that take up all of your time. So many of the quitting stories I heard involved people explaining how it took time to find another job or to acquire the skills and experiences that would allow them to even begin looking for another job if they were employed. This came up in the first LinkedIn workshop I attended. A woman was taking the class because she was currently a language teacher in a middle school and very much wanted to use her foreign language expertise to get a job working for a nonprofit. She had to figure out how to get a job in this new arena, and she thought LinkedIn would be a good place to start. But as the instructor explained in detail how to use LinkedIn as effectively as possible to find a job, she began to get overwhelmed.

She realized that because of teaching, she didn't have time in her day to find something else. This woman was not alone. Rose, whose story I told earlier in the chapter, explained to me that the workload at her job was cyclical enough that there was only a period of a few months in the summer when she didn't work until late at night. That was enough of a breather for her that she could start looking for a job during those months, but she had to quit her search the moment September rolled around. This was frustrating. She ended up looking for three or four years because she had such a limited amount of time to find something else. Enough job seekers explained to me in convincing detail that looking for a job was a full-time job in itself that it was not surprising to hear how difficult people found it to look for a new job while working. Getting to the point at which you can quit your job for something else is often a luxury.

Sometimes there are seemingly insignificant details about how people in workplaces circulate information that can make it difficult to find time to look for another job. One woman who had recently quit her job so that she could search for a new one explained one of the reasons working and searching was incompatible at her old workplace. It was a small company—only forty to fifty people worked there. And whenever anyone was late or leaving early, they emailed everyone who worked there to announce that they had a doctor's appointment or some similar excuse. She got five to six emails a day explaining someone's work schedule, often from people she might never interact with on a regular basis. When she realized she didn't want to stay, she started wondering how she would be able to manage interviewing for jobs during working hours. How many doctors' appointments could she pretend to have? This company's email practice had unintended consequences—contributing to one person's decision to quit sooner rather than later, and without another job in hand already.

These workplace pressures are one of the reasons recruiters can play such an important role in quitting. If people don't have enough time to look for a job, recruiters can step in, offering a tempting job without the hassle of looking at job ads. When Harald decided to switch companies, a recruiter was crucial. He was unhappy at his workplace, so when the recruiter contacted him, it was good timing. Harald was in the fortunate position of constantly having to screen recruiters, so it is also possible that he wasn't very worried about carving out time to start job searching. He was used to recruiters offering a relatively time-efficient solution to this quandary. It also meant that he was only willing to deal with in-house recruiters. He screened for this by making sure the recruiter was willing to acknowledge what company he or she worked for. One week, a recruiter happened to contact him in a way he was willing to respond to. Here's a snippet of my conversation with Harald:

> **Ilana:** When did you first start getting contacted by recruiters?
>
> **Harald:** Constantly. I mean when you have a resume like mine, recruiters contact me every week. Most of them I just delete.
>
> **Ilana:** How are they contacting you?
>
> **Harald:** Lately most of them come through LinkedIn. I mean I have a profile, I keep it up to date. . . . You know, just in case someone wants to send me something I want to hear.
>
> **Ilana:** So what would it be like to get something that you want to hear?
>
> **Harald:** Um . . . what would it be like to get something I want to hear? Well there aren't very many companies I would've responded to. So first of all, a lot of the people who contact me through LinkedIn are working for recruiting firms and I almost always . . . I shouldn't even say almost. I always delete those. I am not interested in talking to recruiters unless I really want to leave my current job. In this case I was contacted by a recruiter who was an employee of Facebook. So he didn't have . . . he didn't have to hide who he was working for.[11]
>
> **Ilana:** And he contacted you how?
>
> **Harald:** Through LinkedIn. He said . . . what did he say? . . . He said something like, "Oh we're looking for a few people to lead our infrastructure efforts in New York and wanted to see if you might be interested in chatting." For whatever reason I wrote him back, pretty tersely. But the fact that I wrote back at all . . . I said, "Sure. Okay, I'll bite. What do

you have in mind?" I think those were my exact words, actually. So he
wrote me a paragraph or so about what he was looking for.

Recruiters play a central role in a workplace where quitting is
expected, even tacitly required in order to have a career, and yet
everyone is working so hard that finding time to look for a job can
be difficult. Of course, employers don't turn to recruiters for every
job, and for jobs without recruiters, finding the next job that allows
you to quit your current one can be difficult, as many people find.

This discussion of recruiters highlights that there are many steps
to quitting and that many intermediaries can be involved. Yet when
people think about quitting, that isn't the social labor they tend to
focus on. Instead, they often think about the moment in which they
tell their boss that they are quitting, in person or by email or PowerPoint. The actual conversation in which this is announced can be
highly charged, and many people told me about searching the web for
suggestions about how you are supposed to do this. One person who
no longer works at Google told me: "I googled how to quit Google."
Yet while the quitting conversation can be symbolically loaded, and
if done in an unusual way, will get talked about at work for days afterward, actually quitting takes a lot of effort beyond the actual conversation. I have already talked about how time-consuming finding
another job or setting up your own business can be while working
a white-collar job. I want to turn now to the quitting story that first
made me realize that in today's economy hiring can only be understood by taking quitting into account.

Chaia had no intention of quitting her job when, in retrospect,
she started the wheels turning that would lead to her quitting. She
began working at a nonprofit organization soon after graduating
with a master's degree. She loved her job, and loved working for her
boss, which immediately made this a different quitting story than
others I heard. After the first year, her boss encouraged her to explore
other possibilities, as good managers are supposed to do these days.
She said:

When I had asked him about opportunities to grow within the company, he had said that we're certainly going to have you do that as much as we can. But the company is also limited and you should always be talking to people and actually . . . he said a goal was that you should at least meet with one person a week to—he wasn't necessarily saying to look for a job. But he was saying be out there because inherently he thought there might be better things for me. He knew I was interested in something else. And I loved my job.

Chaia took his advice, not perhaps meeting a new person every week, but when she heard Brett, an inspiring speaker at a conference, she decided to contact him and invite him to coffee. And this is the moment when it all began.

For Chaia, every stage of quitting for a new job was filled with ambiguity and the work of interpreting and second-guessing conversations. She had a conversation that she greatly enjoyed with Brett, talking to him about a new company he was founding and ideas he had about how to transform the ways people access information about health care. Yet she didn't hear from him immediately afterward, and she had expected a pleasant, courteous response to her follow-up thank-you email. Perhaps she had misinterpreted how productive the conversation seemed to be? I am not sure that Chaia would have thought twice about what she found to be an unexpected email silence, except that he emailed her a week and a half later. And that email was when the work of interpretation truly began, because he invited her to meet with his cofounder. The invitation was unclear—this might be a job interview or it might not be. It was ambiguous enough that she asked other people to help interpret the emailed invitation. She ended up taking her mother's advice, which was to prepare as though it was a job interview but not to be disappointed if it wasn't. Yet when the meeting actually happened, it was clear that they were offering her a job, although the mixed messages continued. They told her that they definitely wanted to hire her, but it was not the right time in the company's growth to add employee number 3, as they were still waiting to hear about funding.

This put her in a waiting space. She knew that funding might come through at any moment, but she didn't know exactly when. And this ambiguity was causing her problems with her current job. She needed to work simultaneously in two ways—first, as though she was staying at her current job indefinitely, and second, as though she was wrapping projects up and preparing to have someone else take over at a moment's notice. Yet, sooner than she expected, an email came that said they didn't want to wait to find out about funding—they wanted her to start immediately. This created a whole new balancing act: as she waited to figure out how to tell everyone in her office that she was leaving, she tried to determine which of her projects she could wrap up quickly and how to avoid starting new projects at her current company while doing so.

In this instance, Chaia had help with the social labor of quitting from her new boss. Brett knew the founder of the company that she worked for and understood how tricky it might be to appear as though he was poaching employees from that company. So Brett contacted the founder himself to explain why Chaia was leaving and to reassure the founder that this new start-up was not in direct competition but rather could be a potential partner in the future. Chaia thought that it was relatively unusual for her new boss to help her navigate the complicated social labor of disentangling from a workplace while maintaining cordial relationships in that workplace for the future. I haven't come across a similar story myself, although I imagine Chaia isn't the only person this has happened to, especially since a company's relationships to other companies can be at stake when an employee switches jobs.

Chaia then had the complicated dilemma of telling the boss whom she greatly admired that she wanted to leave for another job. She explained her quandary:

> How do I describe why it is that I am leaving and not have it be about—it's a little bit like a break up, this isn't about you, this is about me. And again, this is my boss, not my cofounder, and he and I had this fantastic relationship and they had both nurtured me but my boss in particular.

And he lived in Chicago and I'm in California so I never really see him. And by never, I saw him once every six months and I only worked for the company for a year. So I saw him a few times. So that made it just that much harder—how do you have this conversation? I'm also in a place where you don't have a lot of privacy. We're in like a start-up atmosphere so there's people all around me all the time. So I thought: how do I do this? I don't want people to hear me while I am, you know. So I decided that I would call him and say it—I think I used my check-in. We have a weekly check-in. He's very big on minimal meetings so I had to think about how to respect his time but I knew that I needed to take some time to talk to him. I took a walk and I think I might have stayed home from work that day or at least that morning, I think that's what I did so I could be sure that I would have a little more buffer. I had a lot more buffer and I still took the call outside for whatever reason.

Chaia put a great deal of thought into how and when she would tell her boss that she was leaving. He was in Chicago, which meant that she had to do this in a mediated fashion. She couldn't tell him in person. She discarded the possibility of email, as she believed that email would be too disrespectful a way to let him know. However, she felt uncomfortable simply calling him to tell him, especially at work. Hierarchy structured their relationship too much. A spontaneous phone call would have been odd and out of place, and would have violated his clearly expressed expectations of how she should demonstrate valuing his time. In addition, because the way the office space was laid out at work, she couldn't tell him while she was in the office without announcing it simultaneously to all of her other coworkers. She didn't want her coworkers to find out "accidently" by eavesdropping. She wanted to be able to tell them herself. Yet she couldn't tell them until she told her boss. There is an order to how this information should circulate in the workplace. All these complicated social norms and spatial layouts led her to stay home on the day that she knew she would have a regular conversation with her boss and could let him know her plans.

Yet telling the founder and her boss was only half the work of quitting. All her coworkers needed to know she was leaving, but she also needed to let her friends know in a way that signaled to them that she considered her friends at work to be more than mere coworkers. Chaia ended up delegating who told her coworkers in a way that crops up often in other stories I collected—that is, the boss often makes the announcement in a meeting or by email. In this instance, her boss emailed all her coworkers to let them know that she was leaving. But that meant she had to know when her boss was going to do this so that she could text her friends before the email was sent. However, she didn't want to let them know too far in advance, since work gossip spreads like wildfire. Once she finessed that timing, then there was the slightly awkward happy hour in which all her coworkers and friends at work wished her well.

Yet even that did not end all the effort that goes into quitting. She had to send a letter of resignation to HR, a task that befuddled her at first. She went online to figure out what she was supposed to do—this was a completely unfamiliar genre for her. Chaia explained:

> So I thought, how do I do that? So I asked my mom and then I googled it and I found all these templates. They were like: "Do you want it to be funny? Do you want it to be angry?" There were all these different choices. I read through them and I wanted to share them with friends because I thought they were so funny. The angry ones were hilarious: "I've never enjoyed one day in this office. The happiest I'll ever be is the moment I quit." It was amazing. And I thought: How formal should I be? Is this a time where I need to be personal? Is this a time where I need to be thankful? And my mom just said: "This is a time where you make it very brief. It is just for the records." She said: "It's the meetings and the conversations in which you talk about those other things. This is not that." But I didn't know. I didn't know which was which. For all I know, this was where you document everything that you love about your company. I don't know. The templates helped though a lot because I was able to just, it was something that helped me frame it and make it concise. My mother was like: "No one is going to read this."

And then finally, once everyone at work had been told, and HR had its official, formal, and not very informative letter, there was one last task: making it LinkedIn official. She wondered: "Do I make it LinkedIn official after I have told people or after my week and a half of actually finishing, and I think since I had a new job, it didn't matter that much. But I wanted to be respectful of my current coworkers and so I think I waited until it was my last day."

All of this was just the work of getting a new job almost by accident and then managing how this information should properly be told to everyone at the company. And all of the work I just described does not begin to touch on how Chaia and others have to coordinate and manage the tasks that their job requires of them in such a way that someone else can step into their shoes smoothly and easily. Everyone I spoke to talked at length about all the effort they put in so that their quitting did not affect their coworkers. And while I think Chaia was willing to go into more detail with me than others about how she managed all the different steps because this was her first time quitting a job while simultaneously managing a career, everyone goes through a workplace-specific version of considerable effort when they quit a job.

While Chaia's experiences capture quite a bit of the work that goes into quitting these days, because she wasn't trying very hard to get a new job, she didn't mention some of the effort that other people told me about. For example, Rose, who had so much trouble finding time while working to even look for another job, told me about some of the effort that went into figuring out how to manage any career transition. She wanted to move from being communications manager at her company to being an administrative assistant. She got the title *communications manager* because her company thought that anyone who was from a particular generation automatically would know how to use social media to promote the interests of the organization. Yet she knew that being a communications manager at a company required different skills than what she had taught herself at the nonprofit organization where she worked. She wanted a job that she felt more qualified to do and that could lead to other jobs she might enjoy as well, such as project management. This meant that she wanted to get a job whose title might seem to any employer like a step down from the job she had at the time. She felt that in reality, given how her organization was structured, it was not an actual step down. But it might seem that way on paper. This meant that she had to announce one job title to any potential employer who might come across her web presence, but at the same time any information her company released mentioned a different job title. She worried that anyone searching for her on the internet would see her real title on the organization's website, or that a coworker would ask her why she had the wrong title on her LinkedIn profile. And then there was also the risk that someone would compare her LinkedIn profile with the organization's website and decide she was pulling a fast one.

In this case, the public nature of Rose's profile caused her a conundrum while quitting. She had to represent herself appropriately in terms of what her current company role was, but she also had to think about how this role might be interpreted by prospective employers. Rose's precise dilemma might be specific to her situation, but this kind of work points to some of the effort most people have to put into quitting. They often have to align or explain the specifics of what they do in one job, guessing at how other prospective employers might interpret the written traces of their current work situation. Understanding how your particular circumstances could be perceived from another perspective is a certain kind of work, a social labor that some people manage more successfully than others.

As people are moving more quickly from organization to organization, all the social effort I am describing has come to be more and more necessary, particularly the effort people make to keep relationships smooth with coworkers and bosses in the companies they are leaving. People have always had good reasons to get along with their coworkers, but rapid job transitions mean that there is yet another cost for not getting along with someone. This point was vividly brought home to me by a story Adam told about his relationship over fifteen years with a boss he had problems with. He explained that in the first company where he and Greg worked together, when Greg was promoted to be his boss, Greg managed to take an instant dislike to him. Adam didn't know why this was the case, but Greg decided relatively quickly that he was going to fire Adam and found a way to do so. Adam found another job fairly soon after that and discovered that he liked the new job better. Two jobs later, around five years after Greg first fired him, the company Adam was working for went under. He had to get another job, and he had a very promising set of interviews with one company. He and Matt, the hiring manager, seemed to get along, and Matt started the process of hiring Adam. But from Adam's perspective, it all fell apart mysteriously. The recruiter who contacted him let him know that Matt's boss had blocked the hire, and with a little bit of web searching, Adam was able to discover that this was Greg. Fast-forward a number of years, and a few more jobs, and Adam is working at a well-respected company in the Bay Area when he is told that he needs to get in touch with one of the company's sales representatives in Tucson, a person whose name is awfully familiar. Adam does a little bit of checking and realizes that this is a classic story of reversal of fortunes. Within a few years, and because of bad luck on the job market, Greg was now much lower in the corporate hierarchy than Adam was. Adam's story is a vivid example of how all this turnover in jobs can mean that someone may be your boss in one job, your boss's boss at another company, and then far beneath you in the company hierarchy at a third. Maintaining cordial relationships with everyone becomes even more essential in a context of such fluidity. This makes quitting that much more of a tense situation, a moment of potential conflict and opportunity to provoke ill will in a social milieu in which you never know what your working relationship will be in five to ten years. Although I have to say, in most of the quitting stories I heard, while people worried about quitting, in practice the quitting conversations often went very smoothly.

The Online Work of Quitting

Because quitting is so prevalent, part of the social labor that now accompanies quitting is managing your online presences well in anticipation of quitting. Most people focused on their LinkedIn profiles when they talked to me about this. People worried about revealing that they were looking for a job by updating their LinkedIn profiles too much. They often had two strategies for trying to signal to everyone *but* the people at work that they wanted another job. Some people would try to update their profile slowly over time in the hopes that no one would notice. This was the most common strategy that people told me about, but some people also told me about updating their LinkedIn profile all at once over a weekend. By updating everything online at the same time, they hoped to give their boss and coworkers the impression that this was just a chore that they had finally gotten around to doing.

Of course, coworkers knew that these were the two common strategies, and they would often pay casual attention to these kinds of changes in LinkedIn profiles as indications that a colleague wanted to move elsewhere. But the most common tip-off for coworkers was when someone suddenly began to request to be connected on LinkedIn with their coworkers. This was seen as a clear sign that someone had been successful in their efforts to find another position and was about to quit for another job. It was seen time and time again as the signal that people had realized that their connections to everyone in the office might be about to change, that they were no longer about to be contacting people regularly by email or seeing them in meetings or around the office. LinkedIn for many people was social media that gave second-order information about whether someone was about to leave a company for a new job. The situation of course was different if someone had been fired or laid off. In those moments, people might not update their LinkedIn profiles with new information for a while. When people found their professional situation too uncertain, they tended to avoid changing their LinkedIn profiles, and so in these moments it became a relatively bad barometer for determining someone's current employment.

The new model of employment now presupposes that everyone is going to be changing jobs regularly. Depending on the region where you work in the United States, you are either supposed to be changing jobs every two to three years, five to seven years, or eight to ten years. Ideally, you are changing jobs because you quit. In this sense, quitting is the engine that is supposed to run the contemporary job market, although the reality is something else entirely. Over most people's careers, people are likely to experience all forms of leaving a job—from quitting, to being fired, to being laid off.

What happens when quitting is so central to people's ideas of how jobs function these days? Companies face a new challenge. No longer interested in encouraging people to experience company loyalty, businesses now have to find people filled with "passion" that will lead them to work remarkably long hours, if not many years, at a company without the previous guarantees of stability and promotion. Coworkers start expecting their colleagues to quit routinely, and will sometimes quietly observe people's LinkedIn profiles or other activities for the telltale signs that their departure is imminent. And workers themselves are constantly faced with juggling their obligations to the company they work for and creating their next opportunities—finding time to look for other jobs or developing skills or connections that will allow them to more easily move to the next job.

Quitting itself takes work and social know-how. Quitting involves letting everyone at your company know that you are taking another job in ways that also signal how much you value your relationships to them. You have to let your boss know, as well as your coworkers, in such a way that you can continue having cordial ties in the future. They are the people most likely to help you find the next job you take, because after all, the job you are quitting to take is supposed to be only a temporary stepping-stone in your corporate career.

At the same time, when people are quitting, many feel obligated to anticipate their replacement, to get all their projects and different tasks finished or ready to be taken over by someone else. Because of the nature of many jobs, quitting well is also supposed to involve anticipating your replacement, but what that means depends a lot on the kind of job you have and what it means in that specific workplace to have someone pick up the tasks you have left behind. And now that workplaces are meant to be filled with temporary employees, those organizing the work in the workplace have to think a bit more carefully and strategically about how they allocate tasks to accommodate this turnover. Quitting affects how work is structured within the workplace, although not always successfully—sometimes people's knowledge of how to get things done is missed for months or even years after they are gone.

Starting to see yourself as a business has not only contributed to making quitting repeatedly over the course of your career both anticipated and taken for granted: it has also meant that people will sometimes offer different reasons for quitting than they used to. There are always many complicated reasons for quitting a job. When I talked to people, there was rarely only one reason they left a job. But I did notice that one of the reasons that people mention nowadays for quitting is that their managers or coworkers didn't share the same image of themselves that they had. One person explained to me that she quit in part because, after she transferred to a new division in the company, her coworkers "weren't able to key in to my brand." They kept reacting to her in ways that did not match her own self-image. There are many ways to talk about how you might feel unappreciated at work. Now that the language of personal branding is so pervasive, one of the ways to explain being unappreciated is to say people aren't noticing your branded qualities or that you have to quit in case they tarnish your brand. This metaphor has not only shaped people's practices, but it has also changed how they understand the social problems they encounter.

And what about those left behind? Quitting is a double-edged sword for them as well. On one hand, they now have people in other companies who can attest in hopefully detailed and persuasive ways to what kind of worker they have been in the past. In my descriptive sample, 61 percent of people who thought networking helped them get their next job were helped by people they had worked with previously. And if my analysis is correct, and workplace ties can play a significant role in hiring decisions, then having your coworkers move on is only to your benefit when it comes time for you too to leave. But at the same time, the high turnover, especially at the managerial level, means that you are constantly at risk of losing your supporters within the company. If your work is constantly being overseen by someone new, how do you build a reliable reputation at work? As sociologist Richard Sennett points out, "This managerial revolving door has meant that the steady, self-disciplined worker has lost his audience."[12] And if steady, self-disciplined workers so easily lose their appreciative audience, they may be forced to get outside offers to call attention to their value—the cycle just goes on and on.

Conclusion   We Wanted a Labor Force but Human Beings Came Instead

Chris, an independent contractor in his midfifties, knows a lot about what it means to deal with an unstable job market, especially during those moments when you are between gigs and don't know when you are going to get the next one. There was a period in 2012 where he hadn't had a contracting job for a while, and he had no idea how he was going to pay his rent. He realized he might be able to make his rent for another month, but if he didn't get a job soon, he might be homeless. He decided that he needed to get his body ready for this very likely possibility. "I started to sleep on the floor a few hours each night, as long as I could take it, so I could get used to sleeping on a sidewalk or on the dirt. That's how bad it looked. It just seemed hopeless," Chris said. Out of the blue, a staffing agency based in India contacted him and offered him a contract in the Midwest, giving him enough money to make it through this bad patch. But this stark moment, in which he saw homelessness around the corner, is part and parcel of the downside of careers made up of temporary jobs. Chris responded to this possibility in the way that you are supposed to if you are constantly enhancing yourself. He began to train his body for living on the streets, realizing that he needed to learn how to sleep without a bed. He was determined to be flexible and to adapt to potential new circumstances. Seeing the self as a bundle of skills, in practice, means that for some people enhancing your skills involves training yourself to survive being homeless. This too is a logical outcome of our contemporary employment model.

Throughout this book, I have been describing how people are responding to this new way of thinking about work and what it means to be a worker. In the United States, people are moving away from thinking that when they enter into an employment contract, they are metaphorically renting their capacities to an employer for a bounded period of time. Many people are no longer using a notion of the self-as-rented-property as an underlying metaphor and are starting to think of themselves as though they are a business, although not everyone likes this new metaphor or accepts all its implications. When you switch to thinking about the employment contract as a business-to-business relationship, much changes—how you present yourself as a desirable employee, what it means to be a good employer, what your relationships with your coworkers should be like, the relationship between a job and a career, and how you prepare yourself for the future.

The self-as-business metaphor makes a virtue of flexibility as well as the practical ways people might respond in their daily lives to conditions of instability and insecurity. As Gina Neff points out in *Venture Labor*, the model encourages people to embrace risk as a positive, even sought-out, element of how they individually should craft a career.[1] Each time you switch jobs, you risk. You don't know the amount of time you will have at a job before having to find a new one, and you risk how lucky you will be at getting that job and the next job. And with every job transition, you risk the salary that you might make. If there is a gap between jobs, then some people will find that they no longer experience a reliable, steady, upward trajectory in their salaries as they navigate the contemporary job market.[2] Yet this is what you are now supposed to embrace as liberating.

Chris's experiences cycling between employment and increasing periods of unemployment was a familiar story for me. I interviewed so many people in their late forties to early sixties who had a few permanent jobs early in their careers. But as companies increasingly focused on having a more transient workforce, these white-collar workers found their career trajectories veering from what they first thought their working life would look like. They thought that they might climb the organizational ladder in one or maybe even three companies over the course of their lifetime. Instead, they found that at some point in their mid to late forties, they started having shorter and shorter stints at different companies. The jobs, some would say, would last as long as a project. And as they grew older, the gaps between permanent jobs could start growing longer and longer. They struggled to make do, often using up their savings or selling their homes as they hoped to get the next job. Some started to find consulting jobs in order to make ends meet before landing the hoped-for permanent job, and then found themselves trapped on the consulting track—living only in the gig economy. True, not everyone felt like contracting was plan B, the option they had to take because of bad luck. In their book about contractors, Steve Barley and Gideon Kunda talk about the people they interviewed who actively chose this life.[3] I met these people too, but they weren't the majority of the job seekers I interviewed. Because I was studying people looking for a wide range of types of jobs, instead of studying people who already had good relationships with staffing agencies that provided consultants, I tended to meet people who felt their bad luck had backed them into becoming permanent freelancers.[4] These were people who encountered the self-as-business metaphor as a relatively new model, one they felt they actively had to learn in order to survive in today's workplace, as opposed to the younger people I interviewed, many of whom had grown up with the self-as-business model as their primary way to understand employment.

When you think of the employment contract in a new way, you often have to revisit what counts as moral behavior, since older frameworks offer substantively different answers to questions of moral business practice. People have to decide what it means for a company to behave well under this new framework. Consider the self-as-business model. What does a good company do to help its workers enhance themselves as allied businesses? What are the limits in what a company should do? What counts as exploitation under this new model? Can businesses do things that count as exploitation or bad practices now that might not have been considered problems earlier, or not considered problems for the same reasons (and thus are regulated or resolved differently)? Businesses are certainly deeply concerned that workers' actions both at work and outside of work could threaten the company's brand, a new worry—but this is the tip of the iceberg. And the moral behavior of companies isn't the only issue. Can workers exploit the companies they align with now or behave badly toward them in new ways?

Yet while these two metaphors—the self-as-property and the self-as-business—encourage people to think about employment in different ways, there are still similarities in how the metaphors ask people to think about getting hired. In both cases, the metaphors are focusing on market choices and asking people to operate by a market logic. Deciding whether to rent your capacities is a slightly different question than deciding whether to enter into a business alliance with someone, but in both instances you are expected to make a decision based on the costs and benefits involved in the decision. In addition, both metaphorical contracts presume that people enter into these contracts as equals, and yet this equality doesn't last in practice once you are hired. In most jobs, the moment you are hired, you are in a hierarchical relationship; you are taking orders from a boss.[5] Some aspects of working have changed because of this shift in frameworks, but many aspects have stayed the same.

Avoiding Corporate Nostalgia

I talked to people who were thoughtfully ambivalent about this transition in the metaphors underlying employment. They didn't like their current insecurity, but they pointed out that earlier workplaces weren't ideal either. Before, people often felt trapped in jobs they disliked and confronted with office politics that were alienating and demoralizing. Like many people today, they dealt with companies in which they were constantly encountering sexism and racism. Not everyone had equal opportunities to move into the jobs they wanted or to be promoted or acknowledged for the work that they did well.

However, as anthropologist Karen Ho points out, when you have a corporate ladder that excludes certain groups of people, you also have a structure that you can potentially reform so that these groups will in the future have equal opportunities. When you have no corporate ladder—when all you have is the uncertainty of moving between companies or between freelance jobs—you no longer have a clear structure to target if you want to make a workplace a fairer environment.[6] If there is more gender equality in the US workplace these days than there was thirty years ago, it is in part because corporate structures were stable enough and reformers stayed at companies long enough that specific business practices could be effectively targeted and reformed. Part of what has changed about workplaces today is that there has been a transformation in the kinds of solutions available to solve workplace problems.

I see what people said to me about their preference for the kinds of guarantees and rights people used to have at work as a form of critique, not a form of nostalgia. People didn't necessarily want to return to the way things used to be. When people talked to me nostalgically about how workplaces used to function, it was often because they valued the protections they used to be able to rely on and a system they knew well enough to be able to imagine how to change it for the better.

Many people I spoke to were very unhappy with the contemporary workplace's increasing instability. They worried a great deal about making it financially through the longer and longer dry spells of unemployment between jobs. I talked to a man who was doing reasonably well that year as a consultant, and he began reflecting on what the future would hold for his children. He didn't want them to follow in his footsteps and become a computer programmer, because too many people like him were contingent workers. He wanted them to have their own families and reasoned: "If everybody thinks they can be laid-off in two weeks, how would they feel confident enough to be a parent and know that they've got twenty-one years of consistent investment?"

It is not that the people I spoke to necessarily wanted older forms of work. What many wanted was stability. No matter how many times people are told to embrace being flexible, to embrace risk, in practice many of the people I spoke to did not actually want to live with the downsides of this riskier life. The United States does not have enough safety nets in place to protect you during the moments when life doesn't work out. Because you are supposed to be looking for a new job regularly over the course of a lifetime, the opportunities when you might become dramatically downwardly mobile increase. There are more possible moments in which you have to enhance your skills at surviving on much less money or even living rough.

Changing Notions of What Counts as a Good Employment Relationship

When people are thought of as businesses, significant aspects of the employment relationship change. I have already discussed other ways in which this new metaphor has affected workers. The genre repertoire you use to get a job alters to reflect this understanding as you use resumes, interview answers, and other genres to represent yourself as a bundle of business solutions that can address the hiring company's market-specific temporary needs. Networking has changed—what it means to manage your social relationships so that you can stay employed has shifted. Some people I met are now arguing that you treat the companies you are considering joining in the same way you would treat any other business investment: in terms of the financial and career risk involved in being allied with this company. It is not just that you evaluate jobs differently when you know that your job is temporary—deciding you can put up with some kinds of inconveniences but not others. Instead, you see the job as a short-term investment of time and labor, and the job had better pay off—perhaps by providing you with new skills, new networks, or a new way of framing your work experiences that makes you potentially more desirable for the next job. What if this new framework allows workers to have new expectations of their employers, or can safeguard workers' interests in new ways? If you have this perspective, what are the new kinds of demands that employees could potentially make of employers?

For Tom, this new vision of self-as-business was definitely guiding how he was judging the ways companies treated him and what was appropriate behavior. I first contacted Tom because I heard through the grapevine that he refused to use LinkedIn. I was curious, as I had been doing research for seven months by that point and only came across one other person who was not using LinkedIn (and has since rejoined). We talked about his refusal, and he explained to me that LinkedIn didn't seem to offer enough in return for his data. He clearly saw himself in an exchange relationship with LinkedIn, providing data for it to use and in return having access to the platform. Fair enough, I thought: as far as I can tell, the data scientists at LinkedIn and Facebook whom I have met see the exchange relationship in similar ways. Yet Tom decided that what LinkedIn offered wasn't good enough. It wasn't worth providing the company with his personal data. So I asked him about various other sites that he might use in which the exchange might be more equitable, and he lit up talking about these other sites. For Tom, because he saw himself as a business, and viewed his data as part of his assets, he was ready to see LinkedIn as offering a bad business arrangement, one he didn't want to accept. The self-as-business framework allowed him to see the use of certain platforms as instances of participating in business alliances. Some alliances he was willing to enter into, but not all.

This wasn't his only encounter with a potentially exploitative business arrangement. He typically worked as an independent contractor, and a company asked him to come in for a job interview. When he got there, his interviewer explained that the position was a sweat equity job—Tom wouldn't get a salary, but rather he would get equity in the company in exchange for his labor. "Okay" he replied. "So what is your business model?" His interviewer was surprised and discomforted to be asked this. He refused to answer; employees don't need to know the details of the company's business model, he said. Tom felt that this was wrong; because he was being asked to be an investor in the company—admittedly with his labor instead of with money—he felt should be given the same financial details that any other investor in a company would expect before signing on. It sounded to me like Tom's interviewer was caught between two models: wanting the possible labor arrangements now available but unwilling to adjust whom he told what. The interviewer was not willing to follow through on the implications of this new model of employment, and as a result, Tom wasn't willing to take the job. This is one way in which the self-as-business model offers a new way to talk about what counts as exploitation and as inappropriate behavior—behavior that might not have been an issue decades ago, or would have been a problem for different reasons (perhaps because a couple of decades ago, few people found sweat equity an acceptable arrangement).

But this new model also opens up the possibility that companies can have obligations to their employees that they did not have in the same way before. Since companies often don't offer stable employment, they now provide a temporary venue for people to express their passion and to enhance themselves. Can this look like an obligation that businesses have to their workers? Perhaps—businesses could take seriously what it means to provide workers with the opportunities to enhance themselves. Michael Feher argues that if people are now supposed to see themselves as human capital, there should be a renewed focus on what good investment in people looks like—regardless of whether workers stay at a single company.[7] Should companies now help provide training for an employee's next job? Throughout the twentieth century, companies understood that they had to provide their workers training in order for them to do their job at the company to their best of their ability. Internal training made sense both for the company's immediate interests and for the company's ability to retain a supply of properly trained workers over the life of the company.[8] Now that jobs are so temporary, who is responsible for training workers is a bit more up in the air. Yet some companies are beginning to offer support for workers to train, not for the benefit of the company, but so that workers can pursue their passion, should they discover that working at that company is not their passion. Amazon, for example, in 2012 began to provide training for employees who potentially want radically different jobs. Jeff Bezos explained in his 2014 letter to shareholders: "We pre-pay 95% of tuition for our employees to take courses for in-demand fields, such as airplane mechanic or nursing, regardless of whether the skills are relevant to a career at Amazon. The goal is to enable choice." It makes sense for a company to support its workers learning skills for a completely different career only under the contemporary perspective that people are businesses following their passions in temporary alliances with companies.

This model of self-as-business might give workers some new language to protest business practices that keep them from enhancing themselves or entering into as many business alliances as they would like. For example, just-in-time scheduling in practice is currently preventing retail workers from getting enough hours so that they can earn as much as they would like to in a week. This type of scheduling means that workers only find out that week how many hours they are working and when. They can't expect to have certain hours reliably free, and they need to be available whenever their employer would like them to work. Marc Doussard has found that good workers are rewarded with more hours at work. While white-collar workers might get better pay in end-of-the-year bonuses for seeming passionate, retail workers get more hours in the week. If workers make special requests to have certain hours, Doussard discovered, their managers will often punish them in response, by either giving them fewer hours to work or only assigning them to shifts they find undesirable.[9] In practice, this means that workers have trouble holding two jobs or taking classes to improve themselves, as unpredictable shifts will inevitably conflict with each other or class times. Predictable work hours, in short, are essential for being able to plan for the future—either to make sure you are working enough hours in the week to support yourself or to educate yourself for other types of jobs. Since companies are now insisting that people imagine themselves as businesses, what would happen if workers protested when companies don't allow them to "invest" in themselves or when they are thwarted from having as many business partnerships (that is, jobs) as possible? Perhaps employees should now be able to criticize and change employers' practices when they are prevented from being the best businesses they can be because of their employers' workplace strategies.

Or perhaps companies will be sued for other ways in which they are preventing people from acting as businesses. As Julia Tomassetti, a legal sociologist, explained to me, FedEx has already started to face some of these lawsuits. It is no accident that this is happening to FedEx in particular. FedEx, unlike similar companies such as UPS, has been committed for years to categorizing all its drivers as independent contractors. This, as FedEx explains to anyone it allows to be a driver, is a benefit to the drivers: they will be able to take on as many routes as they can afford and learn useful management skills in the process. When FedEx drivers discover that this doesn't tend to happen in practice, they sometimes sue the company. In *Sanders v. FedEx Ground Package Sys., Inc.*, a driver sued FedEx in a New Mexico court because the driver claimed he was prevented from buying FedEx routes to expand his business, while at the same time he was also prevented from selling his route to another driver. In this case, FedEx was being sued because it was acting too much like a traditional employer, expecting to be able to dictate to its employees which route they should take and how they should perform their jobs on this route, instead of a business in the business-to-business alliance that FedEx suggests it offers by "selling" routes to independent contractors. Here is yet another moment in which workers decide to take on the self-as-business model as fully as possible and end up being frustrated by a corporation's decision to do the same only when the company finds it convenient.

I can't help wondering if companies will start being asked to create diverse workplaces to allow workers to enhance their networks, their bundles of relationships, through a similar extension of the self-as-business logic. There are currently jobs for which you can be hired for your network, not just because you know someone who might help you get that job, but because your range of contacts might be instrumentally beneficial for the company. It is possible that employees will start insisting that companies have an obligation to help them enhance their networks. I began wondering about this when some of the job seekers and job quitters I interviewed talked about their former places of employment as stifling their chances to enhance their social/professional networks. For some people, one of the tacit reasons to take a job at certain companies is the work relationships you are likely to forge while you are there. But perhaps if broad, varied networks become an explicitly positive attribute for a job candidate, then this might be something a company can offer its employees. Companies could start providing opportunities to have everyone at the company broaden their networks through systematically hiring people from varied backgrounds (however the company chooses to define the value of different backgrounds). Part of what could make a company a bad employer under this model is being an employer who limits employees' abilities to develop a network at work. As a consequence, both companies and job applicants might start thinking about the kind of breadth desired from the possible networks that can be formed in workplaces.

One way this could potentially play out is if companies began to decide that they need to hire people with networks that bring a form of diversity and a range of different marketable perspectives to the workplace—perhaps based on gender, ethnicity, or cultural background. Bonnie Urciuoli argues that this is part of the logic shaping how liberal arts colleges imagine creating a diverse campus population—the college is providing students a chance to experience cultural and ethnic diversity that they might not otherwise have experienced. And students are supposed to learn how to be tolerant and communicate with a range of people from different backgrounds—diversity skills that are now frequently described as valuable in the workplace. Of course, as Urciuoli points out, this means that students are sorted into people who can represent diversity to others (so that interacting with them is a learning experience) and people who are supposed to constantly be learning how to deal with diversity.[10] Being someone else's learning experience can quickly become tiresome. In the way that colleges approach this issue, the focus is on the individuals' attributes (for example, are they African American?), not the complex social relationships that different workers might bring with them into the workplace (for example, how do they participate in African American communities?).

This may be a sign that earlier arguments for why it is important to have diverse workplaces are changing, and not for the better. People may begin to find it more persuasive to think about diversity not in terms of gender equity, racial equity, or class mobility, but in terms of enhancing everyone's networks. Framing a need to enhance diversity in this way does not lead to trying to address long-standing injustices of who has access to better jobs. Injustice drops out as a persuasive reason for increasing diversity. Instead people will feel compelled to come up with reasons for why diversity would be good in market terms. After all, under this model, you don't want just to have the broadest possible networks because diverse networks are an unqualified good—not if you are operating under a rubric that insists that you evaluate your practices in terms of good business practices. Thus, one of the consequences of this new metaphor is people feel like they have to talk about how innovation, and especially marketable innovation, comes out of having a diverse range of perspectives at a company, instead of a diverse workplace being good for a company's relationship with local communities or other reasons not based on market incentives.

I have been suggesting that the self-as-business model leads you relatively quickly to think that good companies are companies that help people enhance themselves for the next couple of jobs they might want, either by providing support for people to enhance their skills or by providing environments in which they can enhance their networks. In short, some of what companies are supposed to do under this new model is become good places to leave—to allow

people to cycle in and out of their tenure at the workplace, supposedly improving the business while they are there, and in turn having
the business improve them.

The self-as-business model isn't fully entrenched; it hasn't yet
become the given model in all US institutions and walks of life. This
means, however, that some people are trying to figure out how to
expand the model into places shaped still by older logics. The start-up
companies I describe in chapter 6 are good examples of people's
attempts to transform hiring so that people can more fully represent
themselves as though they are businesses. Admittedly, LinkedIn as
a social media site already does quite a good job of allowing people
to represent themselves as businesses, but you could go further. The
start-up companies I discuss are offering a different take on how to
present yourself as a business than LinkedIn does—they are revis-
ing and transforming ways that technologies can encourage people
to presume this rubric in their social interactions. Because of this
continual expansion, while not everyone has adopted this model as
the lens for viewing their working lives, it is getting harder for Americans to avoid encountering the self-as-business approach in their
daily lives.

Traps of the Self-as-Business Metaphor

This is the new rubric for evaluating whether companies are behaving well toward their employees. And it leaves much to be desired.
When a business fails to sell its products, or to win the next contract, it dissolves. But people? What do you do when people fail to
be successful in being a business for one year, or five years? People
don't dissolve like a business. In addition, markets are meant to be
arbitrary—assigning only transient value, and in the case of hiring,
doing so through often inadequate mechanisms. It is an injustice
to see people only in terms of the value that a market assigns them
through the complicated, limited, irrational, and often arbitrary hiring practices that I have described.

There isn't room in this perspective to argue effectively for a living wage. After all, the salary people get is the best salary they can get under market conditions, and there is no reason within this model to link salaries to an actual living wage. In the contemporary United States, there doesn't seem to be much room to argue for health care, since when you are a business, you need to take care of your own assets, which include your health. There was one group of business-people in my fieldwork who thought differently—the people who were providing online work platforms wanted national health care or a guaranteed basic income for everyone. This is not surprising, since when you are working toward creating a nation of freelancers, you start wondering how to persuade people to choose temporary jobs over permanent ones. Admittedly, earlier proponents of this model in which market logic decides everyone's and everything's value, such as Milton Friedman or Friedrich Hayek, argued that the state should provide access to education and national health care for all its citizens.[11] They reasoned that people should be able to enter into the job market on as equal a footing as possible. If the footing was indeed equal, then the market could determine what they were worth. Even Hayek understood that not everyone has the same opportunities to be successful and that bodies are fragile, and so people are vulnerable in a way that businesses are not. In general, when you limit yourself to viewing the self as a business, you can run into all sorts of problems precisely because of the ways in which people are not in fact businesses: they need health care, food, shelter, education, and so on.

This model of self-as-business also doesn't make room for solidarity. If everyone is a business, then there are no structural divisions between people. Under the previous metaphor, you could have a clear distinction between the people who hire or, metaphorically, rent the worker's labor and the worker. Workers could understand that other workers were in the same structural position they were, and employers could understand that other employers also shared their structural position.[12] It was possible to imagine solidarity with people because some people shared a common relationship to work, but not everyone did. Under this new metaphor, solidarity is harder

to imagine, and indeed, as Kathi Weeks points out, a much more dif-
fuse sense of cooperation is all that you can hope to get.[13] Everyone
is supposedly on equal footing—they are all competitors in a marketplace. Indeed, the metaphor of self-as-business presupposes that
competition is at the core of every relationship.[14] Businesses might
enter into alliances with each other, but only as a temporary arrangement against the backdrop of market competition. These alliances
will end the moment one of them becomes disadvantageous to one
party. While these might be good principles for creating market
conditions, these aren't principles that easily lead to meaningful and
long-standing cooperation. In my chapter on networking, I discuss
how some people try to get around this conundrum by advocating
for a pay-it-forward philosophy toward helping acquaintances looking for a job, although in practice so far this has had limited success.

Finally, how, under this framework, do you keep companies committed to behaving well toward their employees or potential employees? As the model of self-as-business becomes more and more
entrenched, this becomes an ever more pressing concern. Legal regulation has always played an important role—indeed, regulations
are increasingly being asked to substitute for business ethics. What
is fascinating about current employment laws is that the rules governing how employers are supposed to treat employees are shifting
to accommodate this new model, albeit very slowly. For example,
lawsuits for companies in the sharing economy like Uber or Home-
joy are causing lawyers and lawmakers to wonder if there should
be a new way to classify employees—the category of dependent
contractor—in addition to the distinction between employee and
independent contractor. The possibility of a new classification for
workers is a sign that people are reimagining work relationships, and,
in this case, trying to find new language to describe the employment
contract as a specific type of business-to-business arrangement. If you
take this new model seriously, regulation is supposed to play a major
role. In the self-as-business model, every social group and every individual is supposed to be seen as a corporation. Under this rubric,
regulation's major focus is to make sure that these corporations—

both people and companies—are behaving properly toward each other, that any competition which undercuts how markets function is prevented from getting out of control.

Laws or regulations in this perspective are not about protecting people's civil rights. People-as-businesses don't have civil rights—they have business risks. Instead, regulations are supposed to perform an important balancing act by ensuring that no business takes on too much risk or too much responsibility—that the risk and responsibility of all interactions are equitably distributed. A bad employment relationship thus is when one party has taken on too much risk or when one party is undermining the other party instead of enhancing it. This is a very different vision of equitable relationships than was previously held, in part because it ignores the fact that the "businesses" in these relationships are often of very different sizes and types. That is, this view assumes that companies, regardless of their size, are potentially at equivalent risk as a single employee in these contractual relationships. What counts as equitable is very different when you systematically ignore size as you evaluate the distribution of risk and responsibility.

I am an anthropologist, which shapes what I ask and what I learn about hiring in contemporary America. As an anthropologist, I am more interested in figuring out what is happening on a day-to-day level in corporate America than giving standardized advice. But there are still some useful ways to think about jobs and hiring that anthropology can offer. Since I started this project hoping that I could figure out how I could help my undergraduate students, what do I tell them now?

I often start by trying to explain what has changed about hiring and what hasn't. Turning yourself into someone employable nowadays involves embodying older and well-established ways to be a unique individual but with a twist: as a visibly managed self with a brand obligated to use today's media. A job seeker has to be distinctive in the ways that he or she manipulates the forms through which people are supposed to entice potential employers—through a web presence, resume, and performance during job interviews. And you have to make sure that these different genres don't contradict each other. While you are trying to be distinctive, there are limits to how distinctive you can be. You still need to be recognizable to potential employers, engaging with standardized genres such as resumes and LinkedIn profiles enough to be read as a desirable job candidate. In previous decades, job seekers have had to figure out how to be distinctive in standardized and desirable ways, but that isn't what is new about hiring these days. What is new is that now the self is supposed to be treated as a business, and this requires different techniques than people used in the past, including branding your self.

I admit, I always express my skepticism about personal branding. It seems to take too much time, especially to produce an impression that no one I interviewed who hired people seemed to care about. And they didn't care for good reasons—the techniques of personal branding don't include essential skills that would also allow people to do most jobs well. How many jobs require a worker to analyze something complex that changes depending on the context and come up with three or four words that are supposed to capture its essence?

But I do talk about more than why personal branding may be a waste of time. I tell my students to be cautious about putting too much time and energy into cultivating weak ties in the hopes that networking luck will get them a job. I say that if you only have a limited amount of kindness and generosity in your heart, don't use it all up on the strangers sniffing mangoes next to you in a supermarket. Be kind to your coworkers too. Workplace ties are important, and these ties are only increasing in importance as people spend less and less time at a given company. And it isn't just your bosses who need to know that you are reliable and thoughtful. Pulling your weight for your coworkers also matters, because while your next job may not depend on the other people on your team, the job after that may, or the job ten years down the line. Because applicant-tracking systems have become such a frustrating obstacle for job applicants, and because too many people are applying for jobs, it helps to have people at a company who can vouch for how good you are as a worker. And this means that jobs that let you work primarily at home or take you every day to a new location often have a hidden cost that many people don't take into account— forming strong workplace ties becomes that much harder. I encourage my students to think about the problems that today's technologies and today's social practices pose for employers who are trying to find good job applicants, and to evaluate advice about weak ties and workplace ties in light of this.

Because my students often aren't sure about what jobs they want once they graduate, I suggest that they also pay attention to what kinds of lived dilemmas they will have to deal with at different jobs. All jobs have lived dilemmas, but not the same ones. Some lived dilemmas are going to be easier for them to live with than others. For example, if they want to be a magician, they might want to pay attention to the fact that audiences will probably appreciate some simple but showy tricks more than the complicated technical tricks that their fellow magicians will value. So the routines that get them a professional reputation among their colleagues will be different from the routines that get them paid.[15] Or if they want to be a professional organizer, they will have to realize that they are helping their clients find ways to live with tolerable clutter in which they can find everything, not teaching everyone to use the same perfect filing system. They will have to be organized but flexible and will have to decide over and over again in each new situation whether they want to err on the side of systematic organization or flexibility.[16]

I point out to my students that the people who are choosing new hires face their own lived dilemmas at their jobs, and that this is worth remembering as they apply for jobs. It helps to keep in mind that not all recruiters they might encounter have the same incentive schemes for filling job requisitions, and this can help explain recruiters' strategies when dealing with job applicants. And because recruiters, hiring managers, and HR managers all have different lived dilemmas in their jobs, applicants are walking into a complicated social context in which people are coordinating their evaluations of other people, and sometimes in unpredictable ways. As a result, people who are hiring are often more interested in what is going on in the company than in any of the candidates for the position. It helps to know as much about the social dynamics of a workplace as possible. But while this knowledge helps, it is by no means a magic bullet.

And I always conclude by telling my students that hiring practices aren't set in stone. They have changed in the past, and they can change again. Many of the hiring practices that make my students unhappy are ones that those hiring can decide not to do. So I ask them to remember what they dislike and not to do this to other people when they are in a position to hire someone. If they don't like an interview question (for example, what is your greatest weakness?), they shouldn't ask it themselves. If they think that standardized resumes and LinkedIn profiles are not good indicators of what they will be like as employees, they should try to find other ways to evaluate job applicants when they themselves are hiring. Some, but by no means all, of what makes hiring so frustrating are interactions that employers can decide not to do. If someone hates not ever hearing back from a company when they didn't get the job, they should be sure to be gracious and clear about the hiring process to the people who apply when they are the ones helping fill a position. I hope my students, and my readers, will, at the very least, be inspired to avoid repeating the practices they disliked so much as job applicants. But most of all, I hope we will all figure out another way of understanding what it means to be a good employee and a good employer.

Acknowledgments

No book ever truly has a single author. When I read my chapters, 255 I see so many sentences and paragraphs which are translations of questions or comments at the end of a talk, or summaries of conversations with friends where I am quite sure my friends had the idea, not me. From the very beginning, this book has depended on other people's generous insights about their personal experiences and willingness to think with me about hiring in contemporary America. It would have been impossible to write this without so many people's good humor, patience, and wisdom. My regret is that I should thank only my academic connections by name here, and not mention the people I met during fieldwork for reasons of confidentiality—I hope that those I don't mention explicitly understand that I deeply appreciate their help throughout this process.

I owe the genesis of this project to Eden Medina and the students in her 2010 class on computer ethics. Many, many people helped me develop the ideas in this book. John Arthos, Ajay Mehrota, and Susan Williams cautioned me that I needed to historicize more carefully the concept that the self is property. Phaedra Pezzullo asked me when and how the self-as-business metaphor shuts down political conversations. Hui-Hui Tung pointed out that job seekers never see themselves as the center of their network. Misty Jaffe reminded me to track what stances job seekers take in interviews and explore how persuasive they are. Stuart Burrows, ever reliable in this regard, told me just the right articles to read that I would never have come across otherwise. Don Brenneis pointed out that everyone hiring is an expert reader reading with other people's judgment in mind. Eric Wibbels explained to me how to design a spreadsheet to code 380 success stories and talked me through the whole process. And Phil Mirowski had many productive conversations with me over Indian food about how neoliberal market conditions came to be designed in certain ways.

Some people helped me work through the initial framing of the project by generously reading and rereading grant proposals: Elizabeth Dunn, Sara Friedman, and Jane Goodman. Others talked to me throughout my fieldwork and writing as I was developing ideas—it is hard to condense in a few words how important these conversations were for me: Amy Cohen, Rosemary Coombe, Melissa Demian, Teri Silvio, and Bonnie Urciuoli. In turning to the sociology and anthropology of work, I have become a newcomer in a remarkably warm and welcoming intellectual community—I would have started studying work earlier if only I had known that these would be my intellectual kaffeeklatsch: Steve Barley, Jan English-Lueck, Christina Gaarsten, Melissa Gregg, Linus Huang, Carrie Lane, Caitrin Lynch, Ofer Sharone, Claudia Strauss, Julia Tomasetti, and JoAnn Yates.

While I was doing fieldwork, a handful of people were my intellectual oasis, offering me a chance to pause and think about all that I was hearing and observing: Steve Barley, Don Brenneis, Charles Briggs, Melissa Cefkin, Nick Couldry, Mark Granovetter, David Henkin, Miyako Inoue, Lyn Jeffery, Laura Jones, Patrick Larvie, Jean Lave, David Leitner, Mark Mancall, Ray McDermott, and Julian Orr. Others not only offered a similar refuge but also opened up doors for me while I was doing fieldwork: Connie Brooks, Jennifer Cheyer, Jeanine Cowan, Kim Fisher, Lynn Fisher, Amy Galland, Rob Lightfoot, John Schwab, Linda Shortliffe, Marcia Stein, Lisa Stotlar, Ron Visconti, and Bob Withers. And I firmly believe that if there is a god of fieldwork, that deity led me to Al Hulvey, a grandmaster of generosity.

As I was beginning to write, I was invited to give talks, which

ACKNOWLEDGMENTS

gave me welcome deadlines to put words on paper, especially for
the introduction. The audiences' responses strongly influenced how
I turned my talks into chapters. At Indiana University, Denny James

and Edward Brudney invited me to a conference on being a precariat, and that along with the audience at my talk for the Center
for Constitutional Democracy shaped my introduction. My thanks
to Rivka Ribak, Michele Rosenthal, and the Haifa Department of
Communication for inviting me to speak to such a thoughtful and
engaged group of scholars. Eitan Wilf invited me to his department
at the Hebrew University, and this sharp audience showed me some
moments where I could be more persuasive. I also am indebted to
those who attended an IU Labor Studies workshop, those present
at my talk for the Poynter Center for the Study of Ethics and American Institutions, and the DataSociety audience in New York City.
Samantha Herrick invited me to Syracuse and started illuminating
conversations about the medieval resonances of what I was observing. Chris Ball and Alejandro Paz invited me to a conference at Notre
Dame, which began many productive conversations, especially with
Chris Ball the next year at Notre Dame. I also had the chance to
present my research at some of the organizations where I did my
research, Promatch and Career Actions Ministry, and those audiences gave me important insights into how to present this material
outside the academe.

I may have interviewed over 150 people and sat through 53 workshops, but then the haunting question arises: what to do with all
those recordings? My thanks to the graduate students who tamed
the wild mound of words spoken in noisy settings: Feray Baskin,
Iris Bull, Sarah Conrad, Lori Hall-Araujo, Blake Hallinan, Hayley
Handler, Dan Hassoun, and Katherine Johnson.

Matt Tomlinson is always so good-humored about reading any
draft I send him, and he reads it faster than any person should be able
to, making me suspect that he has a secret superpower that allows
him to stop time while he engages with the written word. But for this
book in particular, I owe him a greater debt than usual for pointing
out to me that in *The Simpsons* Mr. Burns says "ahoy-hoy."

Writing sentences is one thing; making sure that they actually capture what you want to say in a way that other people will understand is another. Kelly Finefrock-Creed now knows too well my penchant for comma splices and other writing foibles, and has carefully and methodically spared my readers the same knowledge. My anonymous reviewers thoughtfully intervened to point out ways I could make the book more accessible and speak to broader audiences. I also had invaluable help along the way from Omri Ben-Amos, Lisa Braverman, Claudia Breger, Jeanine Cowan, Nathan Ensmenger, Cynthia Gaete, Paul Manning, Ethan Pollock, and Janelle Taylor. A few friends assigned one or two of my chapters to their classes or let me visit their classes and talk about the book, which generated fantastic feedback for me: Michelle Bigenho, Susan Coutin, Mark Fraley, Hugh Gusterson, Jordan Kraemer, Laura Kunreuther, Norma Mendoza-Denton, Galey Modan, David Sutton, and Susanne Unger. Two people read my full manuscript when I had no idea how to make it better—Todd Sanders and Teri Silvio. Teri Silvio pointed out in thirty seconds flat, as she always does, the logical flaws that I had to fix. And Todd Sanders gave me the courage to call it quits and submit the manuscript to presses.

While I was writing the book, a number of visitors came to IU, and conversations with them led me in new directions or cautioned me from going in others: Marc Doussard, Karen Ho, Lilly Irani, Walter Benn Michaels, Julia Tomasetti, Kathi Weeks, and Noah Zatz. And my thanks to my colleagues at Indiana University who are always willing to talk about the rituals and genres of American work life: Richard Bauman, Hamid Ekbia, Nathan Ensmenger, Paula Girshick, Jane Goodman, David Hakken, Rebecca Lave, Susan Lepselter, Nancy Lightfoot, John McGlothlin, and Susan Seizer. And I am grateful to Nathan Ensmenger, Justin Richland, and Marshall Sahlins who came up with clever, funny titles when I couldn't.

I am as grateful for what I was encouraged to omit as I am grateful to people for their insights on what to include. But for Herbert Hovenkamp, I might have written about the Sherman Act in the introduction—he warned me away from doing so with great patience. And I owe Ellen Konar a writerly debt for encouraging me to write an entire book about our contemporary experiences of capitalism without once using the terms *neoliberalism* or *neoliberal*. Ellen was also the person who early on persuaded me to write a chapter on the dilemmas of resigning from a job because she rightly thought this would prove to be a fruitful extension of my interest in breakups and disconnection.

To work on this project, I received institutional support not only from Indiana University's New Frontiers program and IU's Institute for Advanced Study but also from Stanford's Center for Advanced Study in the Behavioral Sciences (CASBS) and Notre Dame's Institute for Advanced Study. This type of support was also accompanied by intellectual support from all who were at CASBS. It was particularly important for me to have people to chat about ideas with while I was doing fieldwork, and so I am especially grateful to Elizabeth Bruch, Dave Dunning, Sam Fleischacker, Sarah Freedman, Shari Johnson, Ellen Konar, Heather Munroe-Blum, Cynthia Pilch, Ethan Pollock, Byron Reeves, Diane Rezendes, Bruce Schulman, Hui-hui Tung, Erik Wibbels, and Steve Woolgar. At Notre Dame, I want to thank in particular Brad Gregory who outlined my conclusion for me in one remarkably productive conversation when I was stuck. And I also want to thank the fellows at Notre Dame who gave me such useful suggestions on how to revise my conclusion and other sections: Laura Bland, Bjarne Funch, Kevin Grove, Lawrence Hall, David Hart, Mary Keys, Henrike Moll, Jaime Pensado, Aletta Quinn, and Otto Santa Ana.

The first seeds for the analytical motivations for this project were planted when I heard Wendy Brown give a talk. These seeds were nurtured by David Graeber and Lauren Leve, who asked me to write an article on neoliberal agency. I owe David Graeber in general for showing me how important it is to write thoughtfully and entertainingly for the widest public possible about what light anthropological theory can shed on people's everyday dilemmas. While David has been a model for me, Lauren was the first to suggest repeatedly that I could do this too, when I had given her no sign previously that I had this in me.

When I first described to my friends how marvelous my editor Priya Nelson is, they would guess that Priya may have some chocolate in her future. "Chocolate?" I would reply. "I think what she has done can only be properly thanked with a trip to Barcelona." Lately, I have been starting to think that my debt will only be properly satisfied by one of the first commercially available trips to Mars, if she is so inclined.

Lastly, my thanks to David Fisher, who from our very first conversation has been talking to me about hiring processes both etic and emic, and whose counsel (and existence) has shaped all my own job searches since.

Notes

Preface

1. Richard Sennett, *The Culture of the New Capitalism* (New Haven, CT: Yale University Press, 2006).

Introduction

1. C. B. Macpherson, *The Political Theory of Possessive Individualism: Hobbes to Locke* (Oxford: Oxford University Press, 1962).
2. Carrie Lane, *A Company of One: Insecurity, Independence, and the New World of White-Collar Unemployment* (Ithaca, NY: Cornell University Press, 2011).
3. I do want to point out that in my fieldwork, people never openly said that they viewed themselves as though they were rentable property or a business, although their language often implied this. Career counselors and motivational speakers would describe people as a business, but this didn't come up spontaneously in my interviews. If anything, people were more likely to compare themselves explicitly to a commodity—and this was inevitably presented as a complaint about how the hiring market forces people to represent themselves. Ofer Sharone also talks about how Israeli job seekers used the commodity metaphor to describe their job-seeking frustrations in *Flawed System / Flawed Self*. But, even though no one talked about being a business in my conversations with them, I still think it is the underlying metaphor. Thinking about yourself as a commodity doesn't account for the ways in which you are supposed to manage yourself, or, as I will discuss, the ways in which different parts of the self-as-business bundle become important in different contexts or sometimes are in tension with each other.
4. Admittedly, these were mainly people involved in the knowledge economy

job market. I didn't talk to people looking for jobs in government or in academia. When I came across people who might be using a different logic to find a job, they were working class and using well-established techniques from earlier capitalist eras. In fact, they often continued to think of themselves as though they were property.

5. See Richard Handler, *Nationalism and the Politics of Culture in Quebec* (Madison: University of Wisconsin Press, 1988); MacPherson, *Political Theory of Possessive Individualism*; Karen Sykes, ed., "Interrogating Individuals," special issue, *Anthropological Forum* 17, no. 3 (2007).

6. Hannah Arendt, *The Human Condition* (Chicago: University of Chicago Press, 1958).

7. Judith Shklar, *American Citizenship: The Quest for Inclusion* (Cambridge, MA: Harvard University Press, 1991), 45.

8. Macpherson, *Political Theory of Possessive Individualism*.

9. Amy Stanley, *From Bondage to Contract: Wage Labor, Marriage, and the Market in the Age of Slave Emancipation* (Cambridge: Cambridge University Press, 1998), 93.

10. Joseph Stiglitz, *The Roaring Nineties: A New History of the World's Most Prosperous Decade* (New York: W. W. Norton, 2004), 183.

11. Peter Cappelli, *The New Deal at Work: Managing the Market-Driven Workforce* (Boston: Harvard Business School Press, 1999).

12. The Portal-to-Portal Act supported employers' interests; it declared that employers had to start paying employees only after they arrived at their workstation, not when they got to work. Meanwhile, the Fair Labor Standards Act was in favor of employees; it determined that employers were responsible for the time it took employees to change clothes, but this portion of the act was overturned in 2014 in the Supreme Court case *Sandifer v. U.S. Steel Corp.*

13. David Graeber, *Towards an Anthropology of Value* (New York: Palgrave, 2001), 162.

14. Melissa Gregg, *Work's Intimacy* (London: Polity Press, 2011).

15. Lane, *Company of One*, 77.

16. Vicki Smith, *Crossing the Great Divide: Worker Risk and Opportunity in the New Economy* (Ithaca, NY: Cornell University Press / ILR Press, 2001), 144.

17. Joyce Lain Kennedy and Thomas J. Morrow, *Electronic Resume Revolution: Creating a Winning Resume for the New World of Job Seeking* (Indianapolis, IN: Wiley Press, 1995), 53.

18. "Sample Resume for an Office Assistant," Monster.com, downloaded March 22, 2014, http://career-advice.monster.com/resumes-cover-letters/resume-samples/sample-resume-office-assistant-2/article.aspx.

19. Bonnie Urciuoli, "Skills and Selves in the New Workplace," *American Ethnologist* 35 (2008): 212.

20. Indeed, most parts of the self-as-business can be consciously and systematically enhanced—skills, experiences, alliances, and assets. The only part of the new bundle of self that can not be enhanced is qualities, since under this

rubric, qualities form the coherent, authentic core of your self, as I discuss in chapter 1.

21. MBO Partners, *2015 State of Independence in America Report* (Herndon, VA: MBO Partners, 2015).

22. Steve Greenhouse. "Noncompete Clauses Increasingly Pop Up in an Array of Jobs," *New York Times*, June 14, 2014, http://www.nytimes.com/2014/06/09/business/noncompete-clauses-increasingly-pop-up-in-array-of-jobs.html.

23. Jean Lave, afterword to *A World of Work: Imagined Manuals for Real Jobs*, ed. Ilana Gershon (Ithaca, NY: Cornell University Press, 2015), 221–27.

24. My goal was to follow Steve Barley and Gideon Kunda's example in their 2006 book on contractors, *Gurus, Hired Guns, and Warm Bodies*. They viewed contracting as a system and set out to learn the perspective of everyone participating. They figured out how a person's role in the process shaped the ways that person thought the contracting game was played.

Chapter One

1. Susan Coutin, "Re/membering the Nation: Gaps and Reckoning within Biographical Accounts of Salvadoran Émigrés," *Anthropological Quarterly* 84 (2011): 809–34; Bonnie Urciouli, "The Semiotic Production of the Good Student: A Peircean Look at the Commodification of Liberal Arts Education," *Signs and Society* 2 (2014): 56–83. See Susan Coutin for the practical problems of branding a war-torn country such as El Salvador and Bonnie Urciouli for the practical problems of branding a college experience.

2. Robert E. Moore, "From Genericide to Viral Marketing: On 'Brand,'" *Language & Communication* 23 (2003): 342, 340.

3. Celia Lury, *Brands: The Logos of the Global Economy* (London: Routledge, 2004), 75.

4. Bruce Schulman, "Brand Name America: Remaking American Nationhood at the Turn of the Century," in *Making the American Century*, ed. Bruce Schulman (New York: Oxford University Press, 2014), 22.

5. Moore, "From Genericide to Viral Marketing," 332.

6. See Arlie Hochschild, *The Managed Heart: The Commercialization of Human Feeling* (Berkeley: University of California Press, 1983).

7. Hochschild, *The Managed Heart*.

8. Walter Benn Michaels, "An American Tragedy, or the Promise of American Life," *Representations* 25 (1989): 73.

9. Ibid., 83–84.

10. Daniel J. Lair, Katie Sullivan, and George Cheney, "Marketization and the Recasting of the Professional Self: The Rhetoric and Ethics of Personal Branding," *Management Communication Quarterly* 18 (2005): 312 (emphasis in original).

11. Promatch and JVS are community organizations for job seekers.

12. See also Kori Allan, "Skilling the Self: The Communicability of Immigrants as Flexible Labour," in *Language, Migration and Social Inequalities: A Critical Sociolinguistic Perspective on Institutions and Work*, ed. Alexandre Duchene, Melissa Moyer, and Celia Roberts (Bristol, UK: Multilingual Matters, 2013), 56–80.

13. Sonia Livingstone, *Children and the Internet: Great Expectations and Challenging Realities* (Cambridge: Polity Press, 2009), 223.

14. Susan Chritton, "Personal Branding and You," *Huffington Post*, February 20, 2013, http://www.huffingtonpost.com/susan-chritton/personal-brands_b_2729249.html.

15. Susan Philips, "Participant Structures and Communicative Competence: Warm Springs Children in Community and Classroom," in *Functions of Language in the Classroom*, Eds. Courtney Cazden, Vera John, and Dell Hymes (New York: Columbia Teachers Press, 1972), 370–94. See also Erving Goffman, *Forms of Talk* (Oxford: Blackwell, 1981); and Erving Goffman, *Frame Analysis* (New York: Harper and Row, 1974).

16. These students came from three different tribes—the Wasco, the Paiute, and the Warm Springs tribes.

17. Alice Marwick, *Status Update: Celebrity, Publicity, and Branding in the Social Media Age* (New Haven, CT: Yale University Press, 2013), 181–82, 196–97.

18. Carl Elliott, *Better Than Well: American Medicine Meets the American Dream* (New York: W. W. Norton and Company, 2003).

19. Alison Hearn, "'Meat, Mask, Burden': Probing the Contours of the Branded Self," *Journal of Consumer Culture* 8 (2008): 206, 214.

Chapter Two

1. The idea that you can and should practice for an interview has been spreading to other countries. Christena Gaarsten, an anthropologist of Swedish employment practices, told me that fifteen years ago, no one in Sweden prepared answers in advance of a job interview. Job interviews were informal and freewheeling conversations, and it would have been very strange to have well-rehearsed answers ready to go. Nowadays, Swedish job interviews resemble the ritualized US exchanges much more, and so Swedish job candidates now practice for their interviews.

2. I first heard about this format for answering interview questions at Indiana University, and that workshop leader talked about the STAR method—situation, task, action, and result. When I asked career counselors if there was a significant difference between the PSR and STAR methods, no one seemed to think so.

3. Richard Bauman, *A World of Others' Words: Cross-Cultural Perspectives on Intertextuality* (Malden, MA: Wiley-Blackwell, 2004), 6.

4. As a guest speaker in a friend's undergraduate class, I talked about Phil's strategies as an example of how someone can change a conversation's participant structure. One of the students asked me if I meant this literally—if he should take his chair and move to the other side of the table to sit next to the interviewer in a job interview. The first thought that flashed in my head was "Oh thank goodness he asked this question and gave me a chance to say out loud 'No, please, please don't do this.'" This is a good example of why I don't like to give advice about how to get a job. What if he hadn't asked, but decided to go ahead and act on what he thought I was recommending?

5. Wanda J. Orlikowski and JoAnne Yates, "Genre Repertoire: The Structuring of Communicative Practices in Organizations," *Administrative Science Quarterly* 39 (1994): 541–574.

6. See Vicki Smith, *Crossing the Great Divide: Worker Risk and Opportunity in the New Economy* (Ithaca, NY: Cornell University Press / ILR Press, 2001): 141–42.

7. See also Smith, *Crossing the Great Divide*, 145–46.

8. Randall Popken, "The Pedagogical Dissemination of a Genre: The Resume in American Business Discourse Textbooks, 1914–1939." *JAC* 19 (1999): 95. Templates for cover letters were around long before anyone began to publish templates for resumes, which Popken says first appeared in university business discourse textbooks starting around 1914. Popular media articles didn't offer resume templates until the 1940s, and these articles weren't very common until the mid-1960s.

9. See also Peter Cappelli, *Why Good People Can't Get Jobs: The Skills Gap and What Companies Can Do about It* (Philadelphia: Wharton Digital Press, 2012).

10. When doing internet research on companies prior to an interview, job seekers sometimes couldn't figure out a company's internal organization from job titles.

11. I was struck by this discussion of eye contact. In my earlier research on Samoan migrants, I had come across pamphlets produced by the New Zealand government in the 1970s that explained to potential employers that Samoans would avoid eye contact in a job interview because direct eye contact was considered disrespectful in Samoan culture.

12. Having watched Matt do a video job interview in his house, I thought it was more likely that Matt didn't actually have an office with a door that he could close. Not everyone has a door to their workspace at home.

Chapter Three

1. Charles Prosser, *Information Book: Getting a Job* (Bloomington, IL: McKnight and McKnight Publishers, 1936), 9.

2. AI here was also acknowledging a widely held belief that contacting people through LinkedIn is not terribly effective, and that if you want to get a reply to a message, you should email the person.

3. "Job-Hunting Tips for the Redundant Manager," *Management Services* 21, no. 10 (1977): 39.

4. People used the term *informational interviewing* more broadly before Bolles to describe general business conversations in which someone wanted information about what was happening within a firm or to establish professional goodwill. See Willard Sanford and William Yeager. *Effective Business Speaking* (New York: McGraw-Hill, 1960), 318–33.

5. Cover letters, by contrast, are now viewed as so passé that job-seeking organizations no longer hold workshops to teach job seekers how to write one.

6. Like Mark Granovetter's study, this is all self-reported information. Unlike Granovetter, I did not get this information from interviews in which I could ask people to elaborate. Instead, I extracted this information from whatever someone chose to tell in front of an audience of fellow job seekers in the three or five minutes allotted to them.

7. Mark Granovetter, *Getting a Job: A Study of Contacts and Careers* (Chicago: University of Chicago Press, 1974), 140.

8. Ofer Sharone, in an article discussing contemporary American attitudes toward weak ties in job searches, points out that what a weak tie means has changed since Granovetter's study: "In the American context, a referrer's willingness and ability to make a referral is not perceived to require a long-standing or close relationship. In fact, referrals are widely presumed to be obtainable from anyone, whether a preexisting weak tie in Granovetter's (1974) sense, with whom one has not been in touch with for 20 years, or a newly formed tie made at a professional association meeting." Ofer Sharone "Social Capital Activation and Job Searching: Embedding the Use of Weak Ties in the American Institutional Context." *Work and Occupations* 41 (2014): 417.

9. Granovetter, *Getting a Job*, 133.

10. In 2014, when I was halfway through my fieldwork, LinkedIn began to allow you to request a LinkedIn connection solely by virtue of being in the same LinkedIn group—beforehand you had to have another offline basis for requesting a connection.

11. Lydia Saad, "The '40-Hour' Workweek Is Actually Longer—by Seven Hours," August 29, 2014, accessed August 21, 2016, http://www.gallup.com /poll/175286/hour-workweek-actually-longer-seven-hours.aspx.

12. Granovetter, *Getting a Job*, 133.

13. Laura Rivera, *Pedigree: How Elite Students Get Elite Jobs* (Princeton, NJ: Princeton University Press, 2015), 3.

14. Alice Marwick, *Status Update: Celebrity, Publicity, and Branding in the Social Media Age* (New Haven, CT: Yale University Press, 2013).

Chapter Four

1. These guidelines all focus on the person creating the profile, not the people interpreting the profile. What this means in practice is that job seekers are surrounded by advice on how to use LinkedIn, and employers get very little advice on how to read profiles.

2. Benjamin Lee Whorf, *Language, Thought, and Reality: The Selected Writings of Benjamin Lee Whorf* (Cambridge, MA: MIT Press, 1956).

3. When I was learning Samoan for the first time in my twenties, I kept mixing up the exclusive plural for "you" that meant "we but not you the listener" and the more inclusive plural "you" that was supposed to refer to everyone present. I wasn't used to second-person plurals that distinguish between people in this way and made social blunders that other Samoan speakers kindly ignored.

4. Naomi Baron, "Who Sets Email Style: Prescriptivism, Coping Strategies, and Democratizing Communication Access," *The Information Society* 18 (2002): 409.

5. Ammon Shea, *The Phone Book: The Curious History of the Book That Everyone Uses but No One Reads* (New York: Perigree Books, 2010).

6. Claude Fischer, *America Calling: A Social History of the Telephone to 1940* (Berkeley: University of California Press, 1992), 71.

7. Adrian Chen, "Inside Facebook's Outsourced Anti-porn and Gore Brigade, Where 'Camel Toes' Are More Offensive than 'Crushed Heads,'" Gawker, February 16, 2012, accessed August 30, 2016, http://gawker.com/5885714 /inside-facebooks-outsourced-anti-porn-and-gore-brigade-where-camel -toes-are-more-offensive-than-crushed-heads (emphasis in original).

8. This has caused problems for communities in which their online identity is linked to an offline presence that Facebook does not recognize as legitimate, largely because Facebook uses other institutions' verification of an identity as evidence that an offline identity is in fact legitimate. This means in practice that drag queens might have a very vibrant offline identity, but because their offline identity is only sanctioned by their social communities and audiences, and not government institutions, they have trouble getting Facebook to recognize their drag queen names. Valeriya Safronova, "Drag Performers Fight Facebook's 'Real Name' Policy," *New York Times*, September 24, 2014, http://www.nytimes.com/2014/09/25/fashion/drag-performers -fight-facebooks-real-name-policy.html.

9. Somaya Ben Allouch, Nalini Kotamraju, and Kirsten van Wingerden. "Employers' Use of Online Reputation and Social Networking Sites in Job Applicant Screening and Hiring," in *Living Inside Mobile Social Information*, ed. James Katz (Boston: Greyden Press, 2014), 255–56. In their study of HR's and recruiters' use of cybervetting in the Netherlands, Allouch, Kotemraju,

and Wingerden found that those screening were far less convinced that job applicants had that much control over how they were represented online.

10. See also Ofer Sharone, "Double-Edged Exposure: Social Networking Sites, Job Searching, and New Barriers to Employment" (unpublished manuscript).

11. The earliest job ad I could find was placed in the *Boston News Letter* in 1705: "A certain person wants a single able man to drive a team in Boston; If any such will repair to John Campbell Post-master of Boston, they may have Encouragment for that Work." Even then, who was doing the actual hiring is a mystery.

12. When this strategy was mentioned at a workshop, some participants thought that this was a good one only if you were trying to get a high-level executive job at a company. They felt that it was only appropriate for interacting with executive recruiters, and that you should avoid it if you want a midlevel office job.

13. Kristin Burnham, "LinkedIn Quick Tip: How to Discover the Hottest New Job Skills," CIO.com, February 9, 2011, accessed July 12, 2015, http://www.cio.com/article/2411310/careers-staffing/linkedin-quick-tip—how-to-discover-the-hottest-new-job-skills.html.

14. Tony tended to email people to let them know he had endorsed them and to encourage them to endorse him in return. He was the only person who told me about doing this.

15. Avoiding typos whenever possible is common advice, especially because it is an easy enough reason for someone to dismiss you when they are going through hundreds of resumes. While I heard this warning often from career counselors, HR professionals, and recruiters, I did also have one or two recruiters tell me that this is too arbitrary and inconsequential a reason to dismiss a candidate. But one's email domain? Many people didn't care. It is too unpredictable whether this will matter. If evaluators notice, they are as likely to interpret this information as a positive sign ("oh, I have an AOL account too") as they are to interpret it as a negative one.

16. For a similar example, see Allouch, Kotamraju, and Wingerden, "Employers' Use of Online Reputation and Social Networking Sites."

17. Molly Wendell, *The New Job Search: Break All the Rules, Get Connected, and Get Hired Faster for the Money That You're Worth* (North Audley Media, 2009).

18. While LinkedIn allows you to represent some of what makes up the bundle of self-as-business, it doesn't allow you to represent all of it. Your assets aren't visible, nor are your qualities. Unlike Facebook users, LinkedIn users don't represent themselves as consumers, nor do they represent themselves, as Etsy allows, as people able to produce particular products to sell.

Chapter Five

1. Madeline Akrich, "The De-scription of a Technical Object," in *Shaping Technology / Building Society: Studies in Sociotechnical Change*, ed. Wiebe Bijker and John Law (Cambridge, MA: MIT Press, 1992), 205–24.

2. Robert E. Moore, "From Genericide to Viral Marketing: On 'Brand,'" *Language & Communication* 23 (2003): 347.

3. SwoopTalent, "Vetting the Best Candidates with Social Sourcing," *Swoop-Talent Blog*, October 17, 2013, accessed November 10, 2013, http://www.swooptalent.com/#!Vetting-the-Best-Candidates-With-Social-Sourcing/c1yy5/EC443387-C02E-4C69-A52D-9FB8B9C17FB3.

4. This is already taking place in other contexts. For example, people who sign up for an invisible lover on an online service such as Invisible Boyfriend or Invisible Girlfriend communicate with any number of people posing as a single fake boyfriend or girlfriend, although admittedly the person who chooses to have a fake lover does get to choose what the lover will look like. Invisible Boyfriend, accessed January 16, 2015, https://invisibleboyfriend.com.

5. Akrich, "The De-scription of a Technical Object."

Chapter Six

1. Stefan Timmermans, *Postmortem: How Medical Examiners Explain Suspicious Death* (Chicago: University of Chicago Press, 2006), 52–53.

2. Jean Lave, afterword to *A World of Work: Imagined Manuals for Real Jobs*, ed. Ilana Gershon (Ithaca, NY: Cornell University Press, 2015), 224 (emphasis in original).

3. William Finlay and James Coverdill, *Headhunters: Matchmaking in the Labor Market* (Ithaca, NY: Cornell University Press, 2007), 33.

4. Finlay and Coverdill, *Headhunters*, 33.

5. No matter how bad the job market was for many of the people I talked to, there were some people in the knowledge economy who were being contacted weekly and even daily by recruiters. In response, they created more and more barriers to stop this flood of contacts. Having this trick in print might mean those computer engineers will have to develop a different strategy. But who knows if recruiters will have read this far.

6. Brenda Berkelaar and Patrice M. Buzzanell, "Cybervetting, Person-Environment Fit, and Personnel Selection: Employers' Surveillance and Sensemaking of Job Applicants' Online Information," *Journal of Applied Communication Research* 42 (2014): 465–66 (quotation at 465). Here Berkelaar and Buzzanell draw on Society for Human Resource Management,

"SHRM Code of Ethical and Professional Standards in Human Resource Management," 2007, accessed June 1, 2011, http://www.shrm.org/about /Pages/code-of-ethics.aspx.

7. Peter Cappelli, *Why Good People Can't Get Jobs: The Skills Gap and What Companies Can Do about It* (Philadelphia: Wharton Digital Press, 2012).

8. This is not a notion of culture that an anthropologist recognizes. It involves seeing culture and identity as synonymous, which means that the values you ascribe to a person can be the same values that you ascribe to a company—cultures, values, and identity are all the same under this framework. Not so for anthropologists. For an anthropologist, culture is both the assumptions about how social interactions should take place and what happens when they don't occur in the ways expected, combined with the ways in which people organize themselves as families, as political units, and as other groups. For an anthropologist, you don't have a culture; a culture has you.

9. Lave, afterword to *A World of Work*, 222.

10. Emmanuelle Marchal, Kevin Mellet, and Geraldine Rieucau, "Job Board Toolkits: Internet Matchmaking and Changes in Job Advertisements," *Human Relations* 60 (2007): 1099.

Chapter Seven

1. While quitting is a crucial extension of this metaphor, retirement doesn't seem to be. If you are a business, not only are you potentially working day and night, as Melissa Gregg points out in *Work's Intimacy*, but you are also facing a future filled with work or self-enhancement.

2. Linus Huang, "Competing Flexibilities in Software Development: The Dynamics and Transformations of Work in a Silicon Valley Startup" (PhD diss., University of California, Berkeley, 2008).

3. Luc Boltanski and Eva Chiapello, *The Spirit of New Capitalism*, trans. Gregory Elliott (London: Verso Books, 2006), 359.

4. Karen Ho, *Liquidated: Ethnography of Wall Street* (Durham, NC: Duke University Press, 2009), 3.

5. Peter Cappelli, ed., *Employment Relationships: New Models of White-Collar Work* (Cambridge: Cambridge University Press, 2008), 6–7; Conference Board. *HR Executive Review: Implementing the New Employment Contract* (New York: Conference Board, 1997).

6. Dave Marcotte, "The Wage Premium for Job Seniority during the 1980s and Early 1990s," *Industrial Relations* 37 (1998): 419–39.

7. Carrie Lane, *A Company of One: Insecurity, Independence, and the New World of White-Collar Unemployment* (Ithaca, NY: Cornell University Press, 2011), 40.

8. Ofer Sharone, *Flawed System / Flawed Self: Job Searching and Unemployment Experiences* (Chicago: University of Chicago Press, 2013).

9. David Sobel, "I Never Should Have Followed My Dreams," Salon, August 31, 2014, http://www.salon.com/2014/09/01/i_never_should _have_followed_my_dreams/.

10. Many financial instruments geared toward saving for retirement or buying homes don't yet take into account the career transitions and temporary or contract jobs that come with living as though one is a business in today's labor markets.

11. As I described in the previous chapter, recruiters who are working for staffing agencies often won't tell prospective candidates the name of the company for whom they are recruiting.

12. Richard Sennett, *The Culture of the New Capitalism* (New Haven, CT: Yale University Press, 2006), 78.

Conclusion

1. Gina Neff, *Venture Labor: Work and the Burden of Risk in Innovative Industries* (Cambridge, MA: MIT Press, 2012).

2. The AARP surveyed 2,492 older recent job seekers (aged forty-five to seventy) in 2014 and found that 48 percent were earning less in their current job than they were in their previous job. Gary Koenig, Lori Trawinski, and Sara Rix, *The Long Road Back: Struggling to Find Work after Unemployment*, Insight on the Issues 101 (Washington, DC: AARP Public Policy Institute, March 2015). Of course, financial compensation isn't the sole deciding factor for everyone when they make their decisions on job offers. Some fortunate people can afford to choose a lower salary for more flexible hours, less stressful working conditions, or other aspects of the job that enhance their quality of life.

3. Steve Barley and Gideon Kunda, *Gurus, Hired Guns, and Warm Bodies: Itinerant Experiences in a Knowledge Economy* (Princeton, NJ: Princeton University Press, 2006).

4. Economist Guy Standing calls people in this situation *precariats*—people who lack economic or job security. They don't have institutionally supported ways to improve their skills for the next job, and they also don't have collective organizations that allow them to have a say in their working conditions—although the Freelancer's Union is trying to address this. Guy Standing, *The Precariat: The New Dangerous Class* (London: Bloomsbury Academic, 2011).

5. Kathi Weeks, *The Problem with Work: Feminism, Marxism, Antiwork Politics, and Postwork Imaginaries* (Durham, NC: Duke University Press, 2011).

6. Karen Ho, "Corporate Nostalgia? Managerial Capitalism from a Contemporary Perspective," in *Corporations and Citizenship*, ed, Greg Urban (Philadelphia: University of Pennsylvania Press, 2014), 267–88.

7. Michael Feher, "Self-Appreciation; or, The Aspirations of Human Capital," *Public Culture* 21 (2009): 21–41. See also James Ferguson, "The Uses of Neoliberalism," *Antipode* 41 (2009): 166–84, for a discussion of how the rationale of investment in human capital can be used to advocate for a guaranteed basic income.

8. Peter Doeringer and Michael Piore, *Internal Labor Markets and Manpower* (Lexington, MA: Heath Lexington Books, 1971).

9. Marc Doussard, "What Comes after the Minimum Wage? The Struggle to Define Good Jobs after Fordism" (unpublished manuscript).

10. Bonnie Urciuoli, "Entextualizing *Diversity*: Semiotic Incoherence in Institutional Discourse," *Language and Communication* 30 (2010): 48–57.

11. Milton Friedman, "The Role of Government in Education," in *Economics and the Public Interest*, ed. Robert A. Solo (New Brunswick, NJ: Rutgers University Press, 1955); Friedrich Hayek, *The Road to Serfdom: Text and Documents, the Definitive Edition*, ed. Bruce Caldwell (Chicago: University of Chicago Press, 2007), 148–49.

12. Christian Marazzi, *Capital and Affects: The Politics of the Language Economy* (Cambridge, MA: MIT Press, 2011).

13. Weeks, *The Problem with Work*, ch. 2.

14. Wendy Brown, *Undoing the Demos: Neoliberalism's Stealth Revolution* (Cambridge, MA: MIT Press, 2015).

15. Graham Jones with Loïc Marquet, "How to be a Magician in Paris," in *A World of Work: Imagined Manuals for Real Jobs*, ed. Ilana Gershon (Ithaca, NY: Cornell University Press, 2015), 44–57.

16. Carrie Lane, "How to be a Professional Organizer in the United States," in Gershon, *A World of Work*, 129–45.

Bibliography

Akrich, Madeline. "The De-scription of a Technical Object." In *Shaping Technology / Building Society: Studies in Sociotechnical Change*, edited by Wiebe Bijker and John Law, 205–24. Cambridge, MA: MIT Press, 1992.

Allan, Kori. "Skilling the Self: The Communicability of Immigrants as Flexible Labour." In *Language, Migration and Social Inequalities: A Critical Sociolinguistic Perspective on Institutions and Work*, edited by Alexandre Duchene, Melissa Moyer, and Celia Roberts, 56–80. Bristol, UK: Multilingual Matters, 2013.

Allison, Anne. *Precarious Japan*. Durham, NC: Duke University Press, 2013.

Allouch, Somaya Ben, Nalini Kotamraju, and Kirsten van Wingerden. "Employers' Use of Online Reputation and Social Networking Sites in Job Applicant Screening and Hiring." In *Living Inside Mobile Social Information*, edited by James Katz, 247–68. Boston: Greyden Press, 2014.

Arendt, Hannah. *The Human Condition*. Chicago: University of Chicago Press, 1958.

Arvidsson, Adam. "Brands: A Critical Perspective." *Journal of Consumer Culture* 5 (2005): 325–58.

———. *Brands: Meaning and Value in Media Culture*. London: Routledge, 2005.

Bader, Sara. *Strange Red Cow and Other Curious Classified Ads from the Past*. New York: Clarkson Potter Publishers, 2005.

Banet-Weiser, Sarah. *Authentic TM: The Politics of Ambivalence in a Brand Culture*. New York: New York University Press, 2012.

Barley, Steve, and Gideon Kunda. *Gurus, Hired Guns, and Warm Bodies: Itinerant Experiences in a Knowledge Economy*. Princeton, NJ: Princeton University Press, 2006.

Baron, Naomi. "Who Sets Email Style: Prescriptivism, Coping Strategies, and Democratizing Communication Access." *The Information Society* 18 (2002): 403–13.

Bauman, Richard. *A World of Other's Words: Cross-Cultural Perspectives on Intertextuality.* Malden, MA: Wiley-Blackwell, 2004.

Berkelaar, Brenda L. "Cybervetting, Online Information, and Personnel Selection: New Transparency Expectations and the Emergence of a Digital Social Contract." *Management Communication Quarterly* 28 (2014): 479–506.

Berkelaar, Brenda L., and Patrice M. Buzzanell. "Cybervetting, Person-Environment Fit, and Personnel Selection: Employers' Surveillance and Sensemaking of Job Applicants' Online Information." *Journal of Applied Communication Research* 42 (2014): 456–76.

Bolles, Richard. *What Color Is Your Parachute? A Practical Guide for Job-Hunters and Career-Changers.* Berkeley, CA: Ten Speed Press, 1970–2014.

Boltanski, Luc, and Eva Chiapello. *The Spirit of New Capitalism.* Translated by Gregory Elliott. London: Verso Books, 2006.

Brammer, Lawrence, and Frank Humberger. *Outplacement and Inplacement Counseling.* Englewood Cliffs, NJ: Prentice-Hall, 1984.

Brown, Wendy. *Undoing the Demos: Neoliberalism's Stealth Revolution.* Cambridge, MA: MIT Press, 2015.

Burnham, Kristin. "LinkedIn Quick Tip: How to Discover the Hottest New Job Skills," CIO.com, February 9, 2011, accessed July 12, 2015, http://www.cio.com/article/2411310/careers-staffing/linkedin-quick-tip—how-to-discover-the-hottest-new-job-skills.html.

Busch, Lawrence. *Standards: Receipts for Reality.* Cambridge, MA: MIT Press, 2011.

Campbell, Sarah, and Celia Roberts. "Migration, Ethnicity and Competing Discourses in the Job Interview: Synthesizing the Institutional and Personal." *Discourse and Society* 18 (2007): 243–71.

Cappelli, Peter, ed. *Employment Relationships: New Models of White-Collar Work.* Cambridge: Cambridge University Press, 2008.

———. *The New Deal at Work: Managing the Market-Driven Workforce.* Boston: Harvard Business School Press, 1999.

———. *Why Good People Can't Get Jobs: The Skills Gap and What Companies Can Do about It.* Philadelphia: Wharton Digital Press, 2012.

Chen, Adrian. "Inside Facebook's Outsourced Anti-porn and Gore Brigade, Where 'Camel Toes' Are More Offensive than 'Crushed Heads.'" Gawker, February 16, 2012, accessed August 30, 2016, http://gawker.com/5885714/inside-facebooks-outsourced-anti-porn-and-gore-brigade-where-camel-toes-are-more-offensive-than-crushed-heads.

Chritton, Susan. "Personal Branding and You." *Huffington Post*, February 20, 2013. http://www.huffingtonpost.com/susan-chritton/personal-brands_b_2729249.html.

Collins, Jane. "One Big Labor Market: The New Imperialism and Worker Vulnerability." In *Rethinking America: The Imperial Homeland in the 21st Century*, edited by Jeff Maskovsky and Ida Susser, 280–99. Boulder, CO: Paradigm Publishers, 2009.

———. "The Specter of Slavery: Workfare and the Economic Citizenship of Poor Women." In *New Landscapes of Inequality*, edited by Jane Collins, Micaela di Leonardo, and Brett Williams, 131–53. Santa Fe, NM: School for Advanced Research Press, 2008.

Comaroff, Jean, and John Comaroff. *Ethnicity, Inc*. Chicago: University of Chicago Press, 2009.

Conference Board. *HR Executive Review: Implementing the New Employment Contract*. New York: Conference Board, 1997.

Coombe, Rosemary. *The Cultural Life of Intellectual Properties: Authorship, Appropriation, and the Law*. Durham, NC: Duke University Press, 1998.

Cooper, Marianne. *Cut Adrift: Families in Insecure Times*. Berkeley: University of California Press, 2014.

Coutin, Susan. "Re/membering the Nation: Gaps and Reckoning within Biographical Accounts of Salvadoran Émigrés." *Anthropological Quarterly* 84 (2011): 809–34.

Coyle, John F., and Gregg D. Polsky. "Acqui-hiring." *Duke Law Journal* 63 (2013): 281–346.

Cruikshank, Barbara. *The Will to Empower*. Ithaca, NY: Cornell University Press, 1999.

Doeringer, Peter, and Michael Piore. *Internal Labor Markets and Manpower*. Lexington, MA: Heath Lexington Books, 1971.

Dorn, Richard. "Investing in Human Capital: The Origins of Federal Job Training Programs, 1900 to 1945." PhD diss., Ohio State University, 2007.

Doussard, Marc. "What Comes after the Minimum Wage? The Struggle to Define Good Jobs after Fordism." Unpublished manuscript.

Duchêne, Alexandre, and Monica Heller, eds. *Language in Late Capitalism: Pride and Profit*. New York: Routledge, 2012.

Du Gay, Paul. *Consumption and Identity at Work*. London: Sage Publications, 1996.

Elliott, Carl. *Better Than Well: American Medicine Meets the American Dream*. New York: W. W. Norton, 2003.

English-Lueck, J. A. *Cultures@SiliconValley*. Stanford, CA: Stanford University Press, 2002.

Erickson, F., and J. Schultz. *The Counsellor as Gatekeeper: Social Interaction in Interviews*. New York: Academic Press, 1982.

Feher, Michael. "Self-Appreciation; or, The Aspirations of Human Capital." *Public Culture* 21 (2009): 21–41.

Fejes, Andreas. "Discourses on Employability: Constituting the Responsible Citizen." *Studies in Continuing Education* 32 (2010): 89–102.

Ferguson, James. "The Uses of Neoliberalism." *Antipode* 41 (2009): 166–84.

Finlay, William, and James Coverdill. *Headhunters: Matchmaking in the Labor Market*. Ithaca, NY: Cornell University Press, 2007.

Fischer, Claude. *America Calling: A Social History of the Telephone to 1940*. Berkeley: University of California Press, 1992.

Fogde, Marinette. "Governing through Career Coaching: Negotiations of Self-Marketing." *Organization* 18 (2011): 65–82.

Foster, Robert. "Corporations as Partners: 'Connected Capitalism' and the Coca-Cola Company." *Polar: Political and Legal Anthropology Review* 37 (2014): 246–58.

———. "The Work of the New Economy: Consumers, Brands, and Value Creation." *Cultural Anthropology* 22 (2007): 707–31.

Foucault, Michel. *Discipline and Punish: The Birth of the Prison.* New York: Vintage, 1975.

Frank, Scott. "Ready for Your Close-Up? Polyvalent Identity and the Hollywood Headshot." *Visual Anthropology Review* 28 (2012): 179–88.

Freeman, Carla. "The 'Reputation' of Neoliberalism." *American Ethnologist* 34 (2007): 252–67.

Friedman, Milton. "The Role of Government in Education." In *Economics and the Public Interest,* edited by Robert A. Solo. New Brunswick, NJ: Rutgers University Press, 1955.

Ganti, Tejaswini. "Neoliberalism." *Annual Review of Anthropology* 43 (2014): 89–104.

Garg, Rajiv, and Rahul Telang. "To Be or Not to Be Linked on LinkedIn: Online Social Networks and Job Search." Working paper, Carnegie Mellon University, 2012.

Garsten, Christina, and Kerstin Jacobsson, eds. *Learning to Be Employable: New Agendas on Work, Responsibility and Learning in a Globalizing World.* New York: Palgrave Macmillan, 2008.

Garsten, Christina, and Helena Wulff, eds. *New Technologies at Work: People, Screens, and Social Virtuality.* Oxford: Berg, 2003.

Gatta, Mary. *All I Want Is a Job! Unemployed Women Navigating the Public Workforce System.* Stanford, CA: Stanford University Press, 2014.

Gehl, Robert. "Ladders, Samurai, and Blue Collars: Personal Branding in Web 2.0." *First Monday* 16 (2011). http://firstmonday.org/ojs/index.php/fm/article/view/3579/3041.

Gershon, Ilana. "Neoliberal Agency." *Current Anthropology* 52 (2011): 537–55.

———. "Selling Your Self in the United States." *Political and Legal Anthropology Review* 37 (2014): 281–95.

Gitelman, Lisa. *Always Already New: Media, History and the Data Of Culture.* Cambridge, MA: MIT Press, 2006.

Goffman, Erving. *Forms of Talk.* Oxford: Blackwell, 1981.

———. *Frame Analysis.* New York: Harper and Row, 1974.

Graeber, David. *Towards an Anthropology of Value.* New York: Palgrave, 2001.

Granovetter, Mark. *Getting a Job: A Study of Contacts and Careers.* Chicago: University of Chicago Press, 1974.

Gregg, Melissa. *Work's Intimacy.* London: Polity Press, 2011.

Greenhouse, Steve. "Noncompete Clauses Increasingly Pop Up in an Array of

Jobs." *New York Times*, June 14, 2014. http://www.nytimes.com/2014/06/09/business/noncompete-clauses-increasingly-pop-up-in-array-of-jobs.html.

Gumperz, John. "Interviewing in Intercultural Situations." In *Talk at Work*, edited by P. Drew and J. Heritage, 302–27. Cambridge: Cambridge University Press, 1992.

Halford, Susan, and Leonard, Pauline. "New Identities? Professionalism, Managerialism and Construction of Self." In *Professionals and the New Managerialism in the Public Sector*, edited by Mark Exworthy and Susan Halford, 102–20. Buckingham, UK: Open University Press, 1999.

Handler, Richard. *Nationalism and the Politics of Culture in Quebec*. Madison: University of Wisconsin, 1988.

Hayek, Friedrich. *The Road to Serfdom: Text and Documents, the Definitive Edition*. Edited by Bruce Caldwell. Chicago: University of Chicago Press, 2007.

Hearn, Alison. "'Meat, Mask, Burden': Probing the Contours of the Branded Self." *Journal of Consumer Culture* 8 (2008): 197–217.

Hirschman, Albert. *Exit, Voice, and Loyalty: Responses to Declines in Firms, Organizations, and States*. Cambridge, MA: Harvard University Press, 1970.

Ho, Karen. "Corporate Nostalgia? Managerial Capitalism from a Contemporary Perspective." In *Corporations and Citizenship*, edited by Greg Urban, 267–88. Philadelphia: University of Pennsylvania Press, 2014.

———. *Liquidated: Ethnography of Wall Street*. Durham, NC: Duke University Press, 2009.

Hochschild, Arlie. *The Managed Heart: The Commercialization of Human Feeling*. Berkeley: University of California Press, 1983.

———. *The Outsourced Self: Intimate Life in Market Times*. New York: Metropolitan Press, 2012.

Huang, Linus. "Competing Flexibilities in Software Development: The Dynamics and Transformations of Work in a Silicon Valley Startup." PhD diss., University of California, Berkeley, 2008.

Indergaard, Michael. "Retrainers as Labor Market Brokers: Constructing Networks and Narratives in the Detroit Area." *Social Problems* 46 (1999): 67–87.

Inoue, Miyako. "Language and Gender in an Age of Neoliberalism." *Gender and Language* 1 (2006): 79–92.

Jacobs, Elisabeth. "Principles for Reforming Workforce Development and Human Capital Policies in the United States." Governance Studies at Brookings, December 2013.

Jacoby, Sanford. *Employing Bureaucracy: Managers, Unions, and the Transformation of Work in the 20th Century*. Mahwah, NJ: Lawrence Erlbaum Associates, 2004.

"Job-Hunting Tips for the Redundant Manager." *Management Services* 21, no. 10 (1977): 39.

Jones, Graham, with Loïc Marquet. "How to be a Magician in Paris." In *A World of Work: Imagined Manuals for Real Jobs*, edited by Ilana Gershon, 44–57. Ithaca, NY: Cornell University Press, 2015.

Kaufman, Bruce. *Hired Hands or Human Resources? Case Studies of HRM Programs and Practices in Early American Industry*. Ithaca, NY: ILR Press, 2009.

Keane, Webb. "Semiotics and the Social Analysis of Material Things." *Language and Communication* 23 (2003): 409–25.

Kennedy, Joyce Lain, and Thomas J. Morrow. *Electronic Resume Revolution: Creating a Winning Resume for the New World of Job Seeking*. Indianapolis, IN: Wiley Press, 1995.

Kerekes, Julie. "Distrust: A Determining Factor in the Outcomes of Gatekeeping Encounters." In *Misunderstanding in Social Life*, edited by J. House, G. Kasper, and S. Ross, 227–57. Harlow: Pearson Education, 2003.

Kilduff, Edward. *How to Choose and Get a Better Job*. New York: Harper and Brothers Publishers, 1921.

Killoran, John B. "Self-Published Web Resumes: Their Purposes and Their Genre Systems." *Journal of Business and Technical Communication* 20 (2006): 425–59.

Kneese, Tamara, Alex Rosenblat, and danah boyd. "Understanding Fair Labor Practices in a Networked Age." Working paper, Data and Society, 2014.

Koenig, Gary, Lori Trawinski, and Sara Rix. *The Long Road Back: Struggling to Find Work after Unemployment*. Insight on the Issues 101. Washington, DC: AARP Public Policy Institute, March 2015.

Lair, Daniel J., Katie Sullivan, and George Cheney. "Marketization and the Recasting of the Professional Self: The Rhetoric and Ethics of Personal Branding." *Management Communication Quarterly* 18 (2005): 307–43.

Lane, Carrie. *A Company of One: Insecurity, Independence, and the New World of White-Collar Unemployment*. Ithaca, NY: Cornell University Press, 2011.

———"How to be a Professional Organizer in the United States." In *A World of Work: Imagined Manuals for Real Jobs*. Edited by Ilana Gershon, 129–45. Ithaca, NY: Cornell University Press, 2015.

Lave, Jean. Afterword to *A World of Work: Imagined Manuals for Real Jobs*. Edited by Ilana Gershon, 221–27. Ithaca, NY: Cornell University Press, 2015.

Linell, P., and D. Thunquist. "Moving In and Out of Framings: Activity Contexts in Talks with Young Unemployed People within a Training Project." *Journal of Pragmatics* 35 (2003): 409–34.

Livingstone, Sonia. *Children and the Internet: Great Expectations and Challenging Realities*. Cambridge: Polity Press, 2009.

Lobel, Orly. *Talent Wants to Be Free: Why We Should Learn to Love Leaks, Raids, and Free Riding*. New Haven, CT: Yale University Press, 2013.

Locke, John. *Second Treatise of Government*. New York: Hackett Publishing, 1980.

Lury, Celia. *Brands: The Logos of the Global Economy*. London: Routledge, 2004.

Macpherson, C. B. *The Political Theory of Possessive Individualism: Hobbes to Locke*. Oxford: Oxford University Press, 1962.

Manning, Paul. "The Semiotics of Brand." *Annual Review of Anthropology* 39 (2010): 33–49.

———. *Semiotics of Drink and Drinking*. London: Bloomsbury Academic, 2012.

Manovich, Lev. *The Language of New Media*. Cambridge, MA: MIT Press, 2002.

Marazzi, Christian. *Capital and Affects: The Politics of the Language Economy*. Cambridge, MA: MIT Press, 2011.

Marchal, Emmanuelle, Kevin Mellet, and Geraldine Rieucau. "Job Board Toolkits: Internet Matchmaking and Changes in Job Advertisements." *Human Relations* 60 (2007): 1091–1113.

Marcotte, Dave. "The Wage Premium for Job Seniority during the 1980s and Early 1990s." *Industrial Relations* 37 (1998): 419–39.

Marvin, Carolyn. *When Old Technologies Were New: Thinking about Electric Communication in the Nineteenth Century*. Oxford: Oxford University Press, 1988.

Marwick, Alice. *Status Update: Celebrity, Publicity, and Branding in the Social Media Age*. New Haven, CT: Yale University Press, 2013.

MBO Partners. *2015 State of Independence in America Report*. Herndon, VA: MBO Partners, 2015.

McGee, Micki. *Self-Help Inc.: Makeover Culture in America*. Oxford: Oxford University Press, 2005.

Michaels, Walter Benn. "An American Tragedy, or the Promise of American Life." *Representations* 25 (1989): 71–98.

Monster Worldwide, "Sample Resume for an Office Assistant," Monster.com, downloaded March 22, 2014, http://career-advice.monster.com/resumes-cover-letters/resume-samples/sample-resume-office-assistant-2/article.aspx.

Moore, Robert E. "From Genericide to Viral Marketing: On 'Brand.'" *Language & Communication* 23 (2003): 331–57.

Nakassis, Constantine. "Brand, Citationality, Performativity." *American Anthropologist* 114 (2010): 624–38.

Neff, Gina. *Venture Labor: Work and the Burden of Risk in Innovative Industries*. Cambridge, MA: MIT Press, 2012.

Newman, Katherine, ed. *Laid Off, Laid Low: Political and Economic Consequences of Employment Insecurity*. New York: Columbia University Press, 2008.

Orlikowski, Wanda J., and JoAnne Yates. "Genre Repertoire: The Structuring of Communicative Practices in Organizations." *Administrative Science Quarterly* 39 (1994): 541–74.

Ottinger, Gwen. *Refining Expertise: How Responsible Engineers Subvert Environmental Justice Challenges*. New York: New York University Press, 2013.

Peters, John Durham. *Speaking into the Air: A History of the Idea of Communication*. Chicago: University of Chicago Press, 1999.

Peters, Rebecca Warne. "Development Mobilities: Identity and Authority in an Angolan Development Programme." *Journal of Ethnic and Migration Studies* 39 (2013): 277–93.

Peters, Tom. "The Brand Called You." *Fast Company*, August 31, 1997. http://www.fastcompany.com/28905/brand-called-you.

Philips, Susan. "Participant Structures and Communicative Competence: Warm Springs Children in Community and Classroom." In *Functions of Language in the Classroom*, edited by Courtney Cazden, Vera John, and Dell Hymes, 370–94. New York: Columbia Teachers Press, 1972.

Piore, Michael, and Charles Sabel. *The Second Industrial Divide: Possibilities for Prosperity*. New York: Basic Books, 1986.

Popken, Randall. "The Pedagogical Dissemination of a Genre: The Resume in American Business Discourse Textbooks, 1914–1939." *JAC* 19 (1999): 91–116.

Prosser, Charles. *Information Book: Getting a Job*. Bloomington, IL: McKnight and McKnight Publishers, 1936.

Puckett, Cassidy, and Eszter Hargittai. "From Dot-Edu to Dot-Com: Predictors of College Students' Job and Career Information Seeking Online." *Sociological Focus* 45 (2012): 85–102.

Rees, Albert. "Labor Economics: Effects of More Knowledge Information Networks in Labor Markets." *American Economic Review* 56 (1966): 559–66.

Rivera, Laura. *Pedigree: How Elite Students Get Elite Jobs*. Princeton, NJ: Princeton University Press, 2015.

Roberts, Celia. *The Interview Game*. London: BBC, 1985.

Roberts, Celia, and S. Campbell. "Fitting Stories into Boxes: Rhetorical and Textual Constraints on Candidate's Performances in British Job Interviews." *Journal of Applied Linguistics* 2 (2005): 45–73.

Rose, Nikolas. *Governing the Soul: The Shaping of the Private Self*. London: Routledge, 1990.

———. *Inventing Our Selves: Psychology, Power, and Personhood*. Cambridge: Cambridge University Press, 1996.

Saad, Lydia. "The '40-Hour' Workweek Is Actually Longer—by Seven Hours," August 29, 2014, accessed August 21, 2016, http://www.gallup.com/poll/175286/hour-workweek-actually-longer-seven-hours.aspx.

Safronova, Valeriya. "Drag Performers Fight Facebook's 'Real Name' Policy," *New York Times*, September 24, 2014, http://www.nytimes.com/2014/09/25/fashion/drag-performers-fight-facebooks-real-name-policy.html.

Sanford, Willard, and William Yeager. *Effective Business Speaking*. New York: McGraw-Hill, 1960.

Scheuer, J. "Recontextualization and Communicative Styles in Job Interviews." *Discourse Studies* 3 (2001): 223–48.

Schulman, Bruce. "Brand Name America: Remaking American Nationhood at the Turn of the Century." In *Making the American Century*, edited by Bruce Schulman, 17–36. New York: Oxford University Press, 2014.

Sennett, Richard. *The Corrosion of Character: The Personal Consequences of Work in the New Economy*. New York: W. W. Norton, 1998.

———. *The Culture of the New Capitalism*. New Haven, CT: Yale University Press, 2006.

Sharone, Ofer. "Double-Edged Exposure: Social Networking Sites, Job Searching, and New Barriers to Employment." Unpublished manuscript.

————. *Flawed System / Flawed Self: Job Searching and Unemployment Experiences.* Chicago: University of Chicago Press, 2013.

————. "Social Capital Activation and Job Searching: Embedding the Use of Weak Ties in the American Institutional Context." *Work and Occupations* 41 (2014): 409–39.

Shea, Ammon. *The Phone Book: The Curious History of the Book That Everyone Uses but No One Reads.* New York: Perigee Books, 2010.

Shidle, Norman. *Finding Your Job: Sound and Practical Business Methods.* New York: Ronald Press, 1922.

Shklar, Judith. *American Citizenship: The Quest for Inclusion.* Cambridge, MA: Harvard University Press, 1991.

Silvio, Teri. "Animation: The New Performance?" *Journal of Linguistic Anthropology* 20 (2010): 422–38.

Skeels, Meredith, and Jonathan Grudin. "When Social Networks Cross Boundaries: A Case Study of Workplace Use of Facebook and LinkedIn." In *GROUP '09: Proceedings of the ACM 2009 International Conference on Supporting Group Work,* 95–104. New York: ACM, 2009.

Smith, Vicki. *Crossing the Great Divide: Worker Risk and Opportunity in the New Economy.* Ithaca, NY: Cornell University Press / ILR Press, 2001.

Smith, Vicki, and Esther Neuwirth. *The Good Temp.* Ithaca, NY: Cornell University Press / ILR Press, 2008.

Sobel, David. "I Never Should Have Followed My Dreams." *Salon,* August 31, 2014. http://www.salon.com/2014/09/01/i_never_should_have_followed_my _dreams/.

Sprague, Robert. "Invasion of the Social Networks: Blurring the Line between Personal Life and the Employment Relationship." *University of Louisville Law Review* 50 (2011): 1–34.

Standing, Guy. *The Precariat: The New Dangerous Class.* London: Bloomsbury Academic, 2011.

Stanley, Amy. *From Bondage to Contract: Wage Labor, Marriage, and the Market in the Age of Slave Emancipation.* Cambridge: Cambridge University Press, 1998.

Star, Susan Leigh, and Martha Lampland, eds. *Standards and Their Stories.* Ithaca, NY: Cornell University Press, 2009.

Stiglitz, Joseph. *The Roaring Nineties: A New History of the World's Most Prosperous Decade.* New York: W. W. Norton, 2004.

Stone, Allucquère Rosanne. *The War of Desire and Technology at the Close of the Mechanical Age.* Cambridge, MA: MIT Press, 1995.

Suchman, Lucy, and Libby Bishop. "Problematizing 'Innovation' as Critical Project." *Technology Analysis and Strategic Management* 12 (2000): 327–33.

Swan, Elaine, and Stephen Fox. "Becoming Flexible: Self-Flexibility and Its Pedagogies." *British Journal of Management* 20: S149–S159.

SwoopTalent. "Vetting the Best Candidates with Social Sourcing." *SwoopTalent*

Blog, October 17, 2013, accessed November 10, 2013, http://www.swooptalent .com/#!Vetting-the-Best-Candidates-With-Social-Sourcing/c1yy5/EC443387 -C02E-4C69-A52D-9FB8B9C17FB3.

Sykes, Karen, ed. "Interrogating Individuals." Special issue, *Anthropological Forum* 17, no. 3 (2007).

Timmermans, Stefan. *Postmortem: How Medical Examiners Explain Suspicious Death*. Chicago: University of Chicago Press, 2006.

Townley, Barbara. "Foucault, Power/Knowledge, and Its Relevance for Human Resource Management." *Academy of Management Review* 18 (1993): 518–45.

Urciuoli, Bonnie. "Entextualizing *Diversity*: Semiotic Incoherence in Institutional Discourse." *Language and Communication* 30 (2010): 48–57.

———. "The Semiotic Production of the Good Student: A Peircean Look at the Commodification of Liberal Arts Education." *Signs and Society* 2 (2014): 56–83.

———. "Skills and Selves in the New Workplace." *American Ethnologist* 35 (2008): 211–28.

Vallas, Steve, and Emily Cummins. "Personal Branding and Identity Norms in the Popular Business Press: Enterprise Culture in an Age of Precarity." *Organization Studies* 36 (2015): 293–319.

Vitak, Jessica, Cliff Lampe, Rebecca Gray, and Nicole Ellison. "'Why Won't You Be My Facebook Friend?': Strategies for Managing Context Collapse in the Workplace." In *iConference'12: Proceedings of the 2012 iConference*, 555–57. New York: ACM, 2012.

Weeks, Kathi. *The Problem with Work: Feminism, Marxism, Antiwork Politics, and Postwork Imaginaries*. Durham, NC: Duke University Press, 2011.

Wendell, Molly. *The New Job Search: Break All the Rules, Get Connected, and Get Hired Faster for the Money That You're Worth*. North Audley Media, 2009.

Whorf, Benjamin Lee. *Language, Thought, and Reality: The Selected Writings of Benjamin Lee Whorf*. Cambridge, MA: MIT Press, 1956.

Index